# Nature and the Environment in Nineteenth-Century Ireland

The Society for the Study of Nineteenth-Century Ireland

The Society for the Study of
Nineteenth-Century Ireland

The purpose of the Society is to promote research into nineteenth-century Ireland, and membership is open to scholars both from Ireland and other countries. The Society welcomes members from a wide range of disciplines: literature, history, economics, geography, sociology, anthropology, theology, women's studies, fine arts, etc. It thus seeks to foster an inter-disciplinary approach to nineteenth-century Irish studies.

Series editors: Laurence M. Geary & Ciara Breathnach

# Nature and the Environment in Nineteenth-Century Ireland

EDITED BY
MATTHEW KELLY

LIVERPOOL UNIVERSITY PRESS

First published 2019 by
Liverpool University Press
4 Cambridge Street
Liverpool
L69 7ZU



British Library Cataloguing-in-Publication data
A British Library CIP record is available

ISBN 978-1-78962-032-0

Typeset by Carnegie Book Production, Lancaster
Printed and bound by CPI Group (UK) Ltd, Croydon, CR0 4YY

# Contents

# Acknowledgements

I am grateful to the Society for the Study of Nineteenth-Century Ireland for inviting me to organise its 2016 annual conference, for being open to the idea that it might be in the environmental humanities, and for including this volume in its series. I would also like to acknowledge the generous support the conference received from the history department at the University of Southampton and the welcome I have since received from my new friends and colleagues at Northumbria University. At Liverpool University Press, Alison Welsby commissioned the volume, Christabel Scaife oversaw the editorial process (including eliciting a very helpful report from an anonymous peer reviewer), and Rachel Chamberlain at Carnegie Book Production helped turn the manuscript into a book. I am grateful to all for their patience and kindness, all proof, if proof were needed, that Liverpool University Press remains a great friend of Irish history. Finally, I would like to thank the contributors for their terrific essays.

M.K.
Riding Mill, August 2019

# Figures

# Tables

# Contributors

**Juliana Adelman** is Lecturer in History and Geography at Dublin City University.

**David Brown** is Professor of Modern History at the University of Southampton.

**Ronan Foley** is Associate Professor in Geography at Maynooth University.

**Huston Gilmore** completed his D.Phil. in Modern History at the University of Oxford.

**Seán Hewitt** is a Leverhulme Trust Research Fellow at Trinity College Dublin.

**Matthew Kelly** is Professor of Modern History at Northumbria University.

**Patrick Maume** works for the *Dictionary of Irish Biography*.

**Helen O'Connell** is Associate Professor of English at Durham University.

**Mary Orr** is the Buchanan Chair of French at the University of St Andrews.

**Colin Reid** is Lecturer in Modern British and Irish History at the University of Sheffield.

# Introduction

## *Matthew Kelly*

To give a collection of essays the title *Nature and the Environment in Nineteenth-Century Ireland* raises tricky conceptual questions that must be confronted. As a string of recent publications make clear, scholars in the environmental humanities find the notion of 'nature' as a referent for an objective material phenomenon problematic. In some respects, this is nothing new. In *Keywords*, Raymond Williams famously stated that 'nature is perhaps the most complex word in the language'. He distinguished three general but overlapping usages, namely 'the essential quality and character of something', the 'inherent force which directs either the world or human beings or both', and the 'material world itself, taking as including or not including human beings'.[1] As he developed his argument, Williams made it clear that the usage of the term 'nature' needs to be historicized, perhaps more pressingly than any other.

Twenty years after Williams's intervention, Peter Coates undertook this exercise in historicization, developing Williams's thinking into a book-length study.[2] As befits his specialism, Coates gave particular space to American writers, concluding with a discussion of Bill McKibben's *The End of Nature* (1990), probably the most influential work of environmental history written in recent times.[3] McKibben's thesis, in Coates's summary, is that 'we have so thoroughly domesticated the earth and modified natural processes that it is no longer possible to speak of nature as something with a separate existence'. Writing about the ecological crisis caused by damage to the ozone layer

1 R. Williams, *Keywords: A Vocabulary of Culture and Society* (New York, 1983), p. 219.
2 P. Coates, *Nature: Western Attitudes since Ancient Times* (Berkeley and Los Angeles, CA, 1998).
3 Incidentally, environmental histories of North America, and the USA in particular, dominate the field, as is made clear in A. C. Isenberg, ed., *The Oxford Handbook of the Environmental History* (Oxford, 2014).

then uppermost in people's minds, McKibben argued that 'by changing the weather, we make every spot on earth man-made and artificial'.[4]

Given the more recent but widespread acceptance of the claim that we inhabit the Anthropocene, a geological epoch formed by the agency of human beings and galvanized by the carbon economy, McKibben's arguments seem at once prophetic and relatively muted.[5] His influence, however, cannot be doubted. A host of academic works have heralded 'nature's end' or positioned themselves as 'after nature', although the response by historians is often less rigidly materialist than this language suggests.[6] The environmental historians Sverker Sörlin and Paul Warde, for instance, reify nature as a phenomenon of which human beings are a component part but not the determining factor. Nature exists without us, and, if it comes to it, will do so after us. The historical subject, they argue, at least in its material aspect, is not nature but the environment, 'a human product, an alloy of nature', constantly re-produced by human labour;[7] leftist readings of the environment, often invoking labour history, identify these environments as 'workscapes', 'material environments made by the continuous confluence of human labour and natural processes'.[8] The notion that environments are constantly re-produced by human labour is easily grasped if we consider what happens to any environment when human labour is withdrawn.[9] Nature makes a comeback, adapting to new conditions, producing a flora and fauna assemblage more natural than its immediate predecessor but still bearing the limiting, conditioning imprint of prior human activities and the continuing, inescapable effects of anthropogenic climate change.[10]

Although the language of production is key to this turn in the literature, it is not of course novel to claim that most rural environments are produced by human activities. For millennia, human beings have created habitats in which non-human nature found a home and often thrives; indeed, nature conservationists have long recognized that the preservation of favoured flora and fauna

4 Coates, *Nature*, p. 176.
5 C. Bonneuil and J.-B. Fressoz, *The Shock of the Anthropocene: The Earth, History and Us*, trans. D. Fernbach (London, 2016), pp. 16–28.
6 J. Lorimer, *Wildlife in the Anthropocene: Conservation after Nature* (Minneapolis, MN, 2015); J. Purdy, *After Nature: A Politics for the Anthropocene* (Cambridge, MA, 2015); S. Sörlin and P. Warde, eds., *Nature's End: History and the Environment* (Basingstoke, 2009); L. Garforth, *Green Utopias: Environmental Hope before and after Nature* (London, 2018).
7 Sörlin and Warde, *Nature's End*, 3.
8 T. G. Andrews, 'Work, Nature, and History: A Single Question, that Once Moved Like Light', in Isenberg, *Oxford Handbook of Environmental History*, p. 426.
9 A. Weisman, *The World Without Us* (New York, 2007).
10 The revival of European wolf populations is partly attributed to the abandonment of uneconomic farms—or workscapes.

communities often requires the maintenance of the human land uses that gave rise to them. When human and non-human nature are perceived successfully to cohabit, creating an apparent harmony between human and non-human nature, the resulting ecosystems are often valorized as 'cultural landscapes' sustained by 'traditional' agricultural or agroforestry uses. Cultural landscapes are often regarded as the obverse of the degraded ecosystems produced by modern industrial and agricultural techniques; defences of certain landscape use as 'traditional', although often playing fast and loose with claims to historical continuity, are also projected as pre-modern survivals. Whether or not we accept these claims and definitions, disrupting this harmony has a catastrophic effect on both nature and human well-being, constituting the 'long emergencies' and enacting the 'slow violence' diagnosed by Rob Nixon.[11]

W. G. Hoskins's seminal *The Making of the English Landscape* (1954), a work whose influence far exceeds its narrow geographical boundaries, was predicated on this view. Hoskins lamented modern developments, finding in the medieval and early modern past a happy balance between people and the landscape. He has since come under stinging criticism from Matthew Johnson, who argues that Hoskins's reading of the history of the English landscape was not only highly selective but that this selectivity rather crudely reflected his own preferences. Still, the notion that landscape is 'made' retains its power.[12] A Hoskinite nostalgia for a pre-modern past found expression in F. H. A. Aalen's historical account of the history of the Irish landscape, in which the 'cultural landscape' of the pre-modern period is posited as embodying a harmony of people and landscape.[13] Given Ireland's early modern history, this seems an extraordinary claim to make, but it reflects less a commentary on history and more a tendency to naturalize pre-modern human–nature relations. It is a position that chimes with claims that the demographic explosion that occurred in Ireland from the middle of the eighteenth century and led to overdependence on potato cultivation had a negative impact on Ireland's biodiversity. It is troubling to learn that the shift from an arable to a pastoral agricultural economy in nineteenth-century Ireland, a process accelerated by the Famine, probably saw a significant recovery in biodiversity.[14] Equally so, Kevin Whelan reminds us that the post-Famine, depopulated west of Ireland

11 R. Nixon, *Slow Violence and the Environmentalism of the Poor* (Cambridge, MA, 2011), p. 3.

12 M. Johnson, *Ideas of Landscape* (Oxford, 2007).

13 F. H. A. Aalen, 'The Irish Landscape: Synthesis of Habitat and History', in F. H. A. Aalen, K. Whelan, and M. Stout, eds., *Atlas of the Irish Rural Landscape* (2nd edn., Cork, 2011), p. 6.

14 This claim is made by D. Cabot, *Ireland* (London, 1999), pp. 272, 420, drawing on F. Mitchell and M. Ryan, *Reading the Irish Landscape* (Dublin, 1997). On structural change to the Irish economy, see A. Bielenberg, 'The Irish Economy, 1815–1880: Agricultural

appealed to tourists and cultural nationalists alike, each imagining that its supposedly 'idyllic blending of culture and environment created a harmonious society and ancient landscape'.[15]

As environmental history became established as a distinct field in the 1970s and 1980s, its dominant narrative mode challenged western narratives predicated on progress and modernization. Most influential, perhaps, was the work of Donald Worster, whose histories of the development of the American west were strongly influenced by the Marxist historiography of the 1960s and 1970s but explicitly positioned as response to their failure to expose the environmental degradation wrought by industrial capitalism.[16] In human geography, the work of Don Mitchell took a similar turn, critiquing work which treated landscape 'as already there to be decoded; the *process of their production is rarely enquired after*'.[17] Although an environmentalist outlook is rarely entirely absent from work in the environmental humanities, scholars are now likely to problematize Worster's 'declensionist' narrative. At a basic level, this is because green activism and more rigorous regulation have had some success, and in some parts of the developed world de-industrialization and farmland abandonment have begun to let nature back, allowing for the development of greater biodiversity; but more significant than these limited claims is how nature is conceived. Rather than nature being something apart from the human, on which human beings inflict harm, scholars are now more likely to treat the natural environment as a co-production of human and non-human nature or as a human/non-human hybrid.[18]

Such formulations recognize that human beings always rely on natural processes and that limits are placed on the human capacity to bend nature to its will. To recognize that limitation, it is suggested, is to acknowledge nature's *agency*, perhaps even the capacity of the natural world to resist or co-operate. Unsurprisingly, the notion that non-humans can exercise agency is often greeted with a pronounced scepticism and to try to establish some kind of proportional judgement about the relative effects of human and non-human nature is a sure path to madness. More easily accepted, perhaps, is the designation of non-human nature as actant, thereby distinguishing

Transition, the Communications Revolution and the Limits of Industrialisation', in J. Kelly, ed., *The Cambridge History of Ireland*, III, *1730–1880* (Cambridge, 2018), pp. 180–5.

15 Kevin Whelan, 'The Modern Landscape', in Aalen, Whelan, and Stout, *Atlas of the Irish Rural Landscape*, pp. 99–110.

16 Worster's pioneering work cannot be summarized here. A good starting point is D. Worster, 'Transformations of the Earth: Toward an Agroecological Perspective in History', *Journal of American History*, lxxvi (1990), pp. 1087–106.

17 Quoted in J. Wylie, *Landscape* (Abingdon, 2007), p. 102.

18 P. S. Sutter, 'The World with Us: The State of American Environmental History', *Journal of American History*, c (2013), p. 96.

human and non-human acts ontologically rather than proportionally. 'As soon as we stop taking nonhumans as objects,' argues Bruno Latour, 'as soon as we allow them to enter the collective in the form of new entities with uncertain boundaries, entities that hesitate, quake, and induce perplexity, it is hard not to see that we grant them the designation of actors'. To reject the distinction between actor and object is to make the human and the non-human properly relational. 'Nothing is easier than to lengthen the list of actants,' Latour suggests, 'whereas one could never manage the relationship between objects and social actors, no matter what dialectical pirouettes one thought oneself supple enough to carry out'.[19] Accepting Latour's advice renders human/non-human relations historically legible; and to do so allows the historian's apprehension when facing sweeping conceptual frameworks, that familiar scepticism born of specificity and archival nuance, to be navigated without either sacrificing the integrity of the historical method or overlooking the findings of more theoretically or conceptually engaged disciplines.

Any attempt to decide when the natural environment became a human/non-human hybrid is likely to take us far from the immediate concerns of this volume. In Ireland's case, it was certainly long before the nineteenth century. More immediately pertinent, then, is to ask when the Anthropocene came into being. This is the geological epoch which scientists look likely to declare succeeded the Holocene, the epoch which began some 12,000 years ago at the end of the last Ice Age. As the term suggests, the Anthropocene is an epoch in which changes to the Earth's life-systems are significantly caused by the actions of human beings. Leading contenders for its commencement include disafforestation in pre-history, the switch from the organic to the inorganic economy in the eighteenth and nineteenth centuries, designated by Andres Malm as the triumph of 'fossil capital', or 'the great acceleration' that followed the Second World War, which included nuclear fallout and continues today with the rapid industrialization of the developing world.[20] Whichever view is taken, Anthropocene thinking gives us a new perspective on our historical subjects. They, like us, were living in the Anthropocene, and if we do interpret the Anthropocene as having been brought into being by industrial modernity, our nineteenth-century subjects, as owners of capital, as workers, and as consumers, were among the agents that catalysed its birth.

That most of Ireland did not industrialize as rapidly as the rest of the UK did not exclude it—or, indeed, does not exonerate it—from the environmental

19 B. Latour, *The Politics of Nature: How to Bring the Sciences into Democracy* (Cambridge, MA, 2004), pp. 76–7.

20 E. A. Wrigley, *The Path to Sustained Growth: England's Transition from an Organic Economy to an Industrial Revolution* (Cambridge, 2016); Andreas Malm, *Fossil Capital: The Rise of Steam Power and the Roots of Global Warming* (London, 2015).

degradation caused by the fossil economy. Ireland, too, was a part of the global commodity flows, which include agricultural produce, fuelled by the fossil economy and mobilized by transport systems—the railways, the steamships, and the trams—by which human beings have assaulted the capacity of the natural world to sustain life. Adopting commercially driven pasture-based agriculture in the nineteenth century—what John O'Leary memorably referred to as the 'Lotus-eating occupation of opening and shutting gates'—was not just socially disruptive but, thanks to ungulate metabolism, also increased Ireland's production of greenhouse gas emissions, contributing to global climate change.[21] It is now clear that as human health and longevity improved, so the capacity of the natural world to sustain life in all its variety diminished. Despite its reputation as a place of natural beauty, Ireland was no exception. Thanks to its island geography, historical disafforestation, intensive farming, and weak environmental lobby, Ireland is now exceptionally bereft of wildlife.[22]

*

*Nature in Ireland: A Scientific and Cultural History* was published in 1997. Edited by John Wilson Foster, this compendious volume of essays was principally but not exclusively concerned with histories not of 'nature itself but the study of nature'. As Foster explained, his purpose was to ensure that the 'systematic study of nature is put inside a cultural and intellectual setting broader than the history of scientific disciplines'.[23] In scholarly work, the gap needed to be closed between 'a pre-Enlightenment concept of natural history, implicated in the economics and politics of plantation and development, and the modern concept of natural history as the classification and understanding of nature in its own right'.[24] These remain worthy aims, but twenty years later this weighty volume, impressive in its scope and the quality of its contributions, feels like the first step in an abandoned project just recently revived. As Juliana Adelman and Francis Ludlow noted in 2014, there *is* a thriving literature on the history of natural history, but Irish environmental history remains in its infancy.[25] Things, though, are starting to change. To take some

21 J. Adelman and F. Ludlow, 'The Past, Present and Future of Environmental History in Ireland', *Proceedings of the Royal Irish Academy: Archaeology, Culture, History, Literature*, cxiv (2014), p. 386.

22 P. Fogarty, *Whittled Away: Ireland's Vanishing Nature* (Cork, 2017).

23 J. W. Foster, ed., *Nature in Ireland: A Scientific and Cultural History* (Montreal, 1997), p. ix.

24 Ibid.

25 Adelman and Ludlow, 'The Past, Present and Future of Environmental History in Ireland', p. 385.

prominent examples. The Trinity Centre for Environmental Humanities is firmly established and has particular strengths in marine and aquatic histories; the *Canadian Journal of Irish Studies* recently dedicated an issue to the 'Irish Environmental Humanities';[26] and University College Cork is host to the 'Deep Maps: West Cork Coastal Cultures' project led by Clare Connolly and Rob McAllen and funded by the Irish Research Council. The *New Cambridge History of Ireland*, a multi-volume enterprise published to great acclaim in 2018, could hardly be described as having embraced environmental history, but Volume II does contain a highly suggestive chapter by Francis Ludlow and Arleen Crampsie on the 'Environmental History of Ireland, 1550–1730'.[27] This chapter reflects the growing influence knowledge of climate change and weather data is having on early modern historiography, including the notion that between the late medieval period and the middle of the nineteenth century the world experienced a 'Little Ice Age'.[28] As such, it is all the more striking that despite the political and cultural emphasis currently placed on climate change and global warming, historians of the modern period have been slow to consider the possible causative effect of warming and shifts in weather patterns that have occurred since the middle of nineteenth century. Does this reflect the resilience of the modernist notion that nature has been made subject to human beings? It is interesting to note, though it is not clear whether it was significant, that in Ireland the shift from an arable to a pastoral agricultural economy coincided with warming, or what used to be described as 'climate improvement'.

This volume hopes to be a part of this turn towards the environmental. It has been developed primarily from papers delivered to the Society for the Study of Nineteenth-Century Ireland conference on 'Nature and the Environment in Nineteenth-Century Ireland' at the University of Southampton in 2016. Although the volume is broadly historical, the essays presented here constitute responses by historians, literary critics, and human geographers of Ireland to the environmental turn taken by the humanities. Not all the contributors identify as working in the environmental humanities, and for some this is their first foray into the field, but each essay constitutes a determined attempt to open up fresh perspectives on familiar subjects and identify new areas of research. Each essay makes it abundantly clear that nineteenth-century Ireland provides fertile ground for environmental approaches, demonstrating that relatively

26 D. Gladwin and M. O'Connor, eds., 'Irish Environmental Humanities', *Canadian Journal of Irish Studies*, xl (2017).

27 F. Ludlow and A. Crampsie, 'Environmental History of Ireland, 1550–1730', in J. Ohlmeyer, ed., *The Cambridge History of Ireland*, II, *1550–1730* (Cambridge, 2018), pp. 608–37.

28 G. Parker, *Global Crisis: War, Climate Change and Catastrophe in the Seventeenth Century* (London, 2013).

modest shifts in intellectual orientation can yield exciting results. Moreover, Ciaran Boylan's observation that the Great Famine was an 'ecological shock on an exceptional scale' is a reminder that the most historically significant event to have occurred in nineteenth-century Ireland cannot be grasped without an environmental perspective.[29] To account for this dimension of the Famine does not sideline human agency, if we accept that ecologies, as subsets of the environment, are co-productions of human and non-human nature. Adelman and Ludlow's observation that the Famine could be approached as an example of how societies cope with environmental disaster chimes with the approach taken by the some of the essayists in this volume.[30] Framing the Irish past in terms of the global perspectives provided by environmental humanities is not just a way of guarding against the parochial tendencies of the Irish academy or encouraging non-Irish historians to engage with the subject, but is a means by which the fuller significance of Ireland's past can begin to be grasped.

Helen O'Connell's essay on 'the nature of improvement' in nineteenth-century Ireland analyses how in the first half of the nineteenth century advocates of agricultural improvement in Ireland tended to characterize Ireland as suffering from an 'excess of naturalness'. This was thought to be manifest in customary agrarian practices and continued sectarian conflict and compounded by population growth and the profusion of the cabins of the rural poor, particularly in the west of Ireland. If Ireland were to thrive, it had to become less 'natural' and more 'social' by embracing new agricultural techniques and methods, thereby building a new kind of society, more harmonious and prosperous. Prosperity would bring improved living conditions, generating better-ventilated, more-hygienic living conditions which would have a morally improving effect on the Irish peasantry. An optimistic outlook and a faith in the capacity of the rural Irish to learn and adopt new agricultural processes were central to improvement doctrine, and much emphasis was placed on their agency, particularly the role of women as powerful figures in the household. Similarly, central to improvement doctrine was the belief that the diffusion of useful knowledge was progressive and apolitical, despite the fact that its often-stated purpose was to integrate the Irish into UK-wide food systems, thereby fulfilling the 'assimilationist' ambitions of the Act of the Union.[31]

The paternalistic outlook of the improvers is a component of David Brown's analysis of Lord Palmerston's attempt to 'conquer' Sligo. Riffing on the title of David Blackbourn's influential analysis of state-led efforts to transform

29 C. Boylan, 'The Famine', in R. Bourke and I. McBride, eds., *The Princeton History of Modern Ireland* (Princeton, NJ, 2016), p. 403.

30 Adelman and Ludlow, 'The Past, Present and Future of Environmental History in Ireland', p. 385.

31 K. Theodore Hoppen, *Governing Hibernian: British Politicians and Ireland, 1800–1921* (Oxford, 2016), p. 118.

the German physical environment between the seventeenth and twentieth centuries, Brown considers Palmerston's attempt to deploy improvement doctrines on his Sligo estates in the decades before the Famine.[32] Palmerston's motives were a classic case of enlightened self-interest: improvement would deliver the moral goods promised by the improvers and raise the yield of the estate, allowing its benevolent owner to settle his debts. Building a new transport infrastructure, including roads and a harbour at Mullaghmore, allowing the development of west coast fisheries and a kelp industry as well as possibly becoming a disembarkation point for Quebec timber, was intended to be no less transformative than his plans to drain the bogs and create a new class of smallholders, plant trees and develop a commercial timber sector, and stabilize the coast by planting bent-grass. All of this sought more effectively to mobilize the energy of nature, including that of his tenants. To this day, parts of the Sligo landscape archive Palmerston's successes and failures, however obscured they now are by natural regenerative processes or subsequent usages.

A more troubling consequence of Palmerston's efforts is explored by Brown's assessment of their possible impact on the Famine in Sligo. As Brown argues, it is difficult to establish whether 'improvement' heightened or lessened the impact of the potato blight on Palmerston's tenants, but it is clear that his priority was to increase yields rather than diversify production. Palmerstonian improvement intensified the dependence on potato monoculture and Brown suggests that this might be considered through the optic provided by James Scott's *Seeing Like a State* (1999). Scott's hugely influential examination of why 'certain schemes to improve the human condition have failed' argues that the modern impulse to pursue politically legible projects with clearly quantifiable deliverables—'the administrative ordering of nature and society'—leads to environmental simplification, that in the twentieth century generated junky ecologies dependent on chemical pesticides and fertilizers and prone to disease.[33]

Post-Famine agrarian politics is the subject of an impressive historical literature which can be only briefly summarized here. Broadly speaking, the circumstances of the Irish tenant farmer improved in the post-Famine decades because rents were stagnant and yields improved, placing the landlord in a relatively disadvantaged position. When landlord attempts to recover their position coincided with looming famine conditions in the late 1870s, the Land War (1879–82) was the consequence.[34] This political conflict, galvanizing

32 D. Blackbourn, *The Conquest of Nature: Water, Landscape and the Making of Modern Germany* (London, 2006).

33 J. C. Scott, *Seeing Like a State: How Certain Schemes to Improve the Human Condition Have Failed* (New Haven, CT, 1999), p. 4.

34 In general, see W. E. Vaughan, *Landlords and Tenants in Mid-Victorian Ireland* (Oxford, 1994).

one of the most remarkable popular mobilizations in modern history, was a response to increased rents and eviction rates, the fear of scarcity, and a generation or more of intense political debate about the rights of the tenantry. Of particular importance was whether tenant farmers had the right to a secure tenure, irrespective of market conditions. Much of this agenda reflected a nationalist reading of the Land Question which distinguished a settler landlord class from a native tenancy. Colin Reid, however, suggests Isaac Butt's role in this debate and his critique of political economy can be interpreted less as promoting nationalist interests and more as promoting environmental justice. Did tenant farmers have a right to extract a living from Ireland's natural resources ahead of the landlord's property rights? And by investing their labour in the soil and improving its agricultural yield did they acquire greater possessory right? If so, insecure systems of tenure, which disincentivized improvement, diminished possessory right. Much of the literature on environmental justice examines how the power of the state–capital nexus diminishes indigenous or working-class rights to natural resources by preventing or inhibiting established forms of resource extraction.[35] Isaac Butt's critique of Irish landlordism as an exercise in wealth extraction by a relatively mobile elite backed by the authority of the state chimes with present-day work on access to natural resources and depredations of global capital. For Butt, agricultural improvement was less a means to increase the profits of rent seekers and more a means to keep the Irish population in place by enabling small-scale, profitable farming. If we follow Reid and look at Irish land politics through the lens of environmental justice, we can start to think about it comparatively, fully registering its broader historical and contemporary significance.

Huston Gilmore's analysis of Daniel O'Connell's 'Monster Meetings' suggests a different way in which Irish people interacted with their natural environment, highlighting alternative forms of possessory claim. Place, identity, and a discourse of nationalist grievance were negotiated through a historicist encounter with the Irish landscape. If the gatherings themselves carried a possessory charge, temporarily helping to reconstitute the relationship of the people with a land associated with dispossession and enabling a form of symbolic or temporary repossession, the meetings were nonetheless largely dependent on the generosity or sympathy of landowners. Similarly, O'Connell's rhetoric was unthreatening to the comfortable yeoman and wealthy farmers in his audience. He extolled the beauty of Irish 'Nature', the pleasures of his estate, and deployed an improvement discourse that imagined how Ireland could more effectively exploit its natural resources. At the same time, in the practices of the participants, a kind of charivari occurred, during which norms

35 For a general introduction, see G. Walker, *Environmental Justice: Concepts, Evidence and Politics* (London, 2012).

were briefly subverted and people felt sufficiently emboldened to defy Ireland's established spatial politics and property. The decorative use of flora, especially 'green boughs'—leafy branches—transformed the locations of the monster meetings into Repeal environments. This practice was discouraged by Repeal marshals on the grounds that trees were private property, throwing into sharp relief the complex relationship Repeal activism had with the law and the scarcity of trees, which thanks to extensive disafforestation were largely associated with the demesne lands of landed estates.[36]

Whereas Gilmore shows how the Monster Meetings temporarily reconstituted certain Irish environments as Repeal space, Ronan Foley looks at a very different process of space-making with respect to Ireland's therapeutic landscapes. Despite the endemic poverty, periodic violence, and political conflict of Irish rural life, the selective representation of Irish rurality in the nineteenth century could make all non-urban space a healthy antidote to town and city life. Foley's more specific arguments concern Ireland's commercialized spas and vernacular sweat-houses as outdoor healing sites. Locating his analysis at the interstices of health geography and environmental history, Foley finds common ground in how both disciplines emphasize place-based practice, hybridity, and nature as actant; moreover, both types of site, despite their different underpinnings—the one communal and folkish, the other commercial and scientific—harnessed the health-giving qualities of spring water and its mineral content, making nature and culture co-produce healing through purgative means in ways which blurred the distinction between the inside and the outside. The decline in the use of these types of healthy place, Foley suggests, saw healing orthodoxies denatured, raising intriguing questions about how the current vogue for 'green prescribing' might be historicized.

My own essay raises related questions about access to beautiful landscapes and the role the state should play in delivering these public goods. Since the eighteenth century, the Killarney Lakes had become established as one of the most aesthetically pleasing landscapes in the United Kingdom, a status that gave them transnational significance and generated a sense that they were a form of common possession held in trust on behalf of the people, Irish or otherwise. This was exposed as no more than fanciful rhetoric when the Muckross Estate came on the market in the 1890s. No legal defence of customary rights of access could be mounted. Considerable public pressure to buy the estate on behalf of the people became entangled in a broader set of questions concerning the capacity of the union state—or the readiness of its politicians—to deliver justice for Ireland. Inspiration was taken from the US federal government's decision to designate state-owned National Parks in

36 T. Reeves-Smith, 'The Natural History of Demesnes', in Foster, *Nature in Ireland*, p. 552.

the 1870s. For Irish nationalists, the questions raised by the Muckross issue posited a form of environmental justice predicated on notionally national and transnational rights quite distinct from Isaac Butt's concern for the rights of Ireland's tenantry. By taking the story into the twentieth century, and possibly abusing editor's privilege, I try to demonstrate how the gifting of the Muckross Estate to the Irish Free State in 1932—on the understanding that it would be run as a National Park—led to a disappointing performance that became a potent symbol of economic stagnation and short-sightedness. In the context of the North Atlantic Archipelago, the Irish Free State and the Republic of Ireland can claim some primacy in its possession of a National Park, but in practice it only began to manage the Park according to recognizably international standards with the modernization of the state in the 1960s and 1970s.[37] Only then did it begin to develop the bureaucratic structures and systems of expertise characteristic of the nature state.[38]

Juliana Adelman provides a very different way of exploring a historical conjunction between human and non-human nature. Her subject is Dublin's animal geographies, here analysed through a pioneering exercise in multispecies history or 'more-than-human' history. More than a decade ago, Harriet Ritvo considered the emergence of animal history, proposing that animals are 'the latest beneficiaries of the democratizing tendency within historical studies'. Histories of class, gender, and race are now joined by histories written by advocates of the historical experience of non-human animal life. Even agricultural historians are now more likely to treat animals as 'individuals'.[39] Some might see this as reflecting a 'post-humanist' turn in scholarship,[40] though as Etienne Benson's study of the urbanization of grey squirrels in the USA demonstrates, the mix of human and non-human agency at work in animal history is complex.[41] By showing how the social geographies of Dublin can be mapped through the distribution of its animal-based industries, Adelman demonstrates not only how cities are places of non-human nature, but also how the regulation of animal life is central to the rise of sanitary

37 National Parks in the Republic of Ireland have yet to find their historian, although important work has been done on their ecology, often as a part of continuing attempts by state authorities to nurture their biodiversity. A notable addition to the literature by a historical geographer is A. Latocha, 'Past Human Activities Recorded in the Landscape: A Case Study from the Glenveagh National Park, Ireland', *Landscape Research*, xl (2015), pp. 338–58.

38 See Introduction to W. Graf von Hardenberg, M. Kelly, C. Leal, and E. Wakild, eds., *The Nature State: New Approaches to the History of Conservation* (London, 2017).

39 H. Ritvo, 'Animal Planet', *Environmental History*, ix (2004), p. 205; also H. Ritvo, 'Among Animals', *Environment and History*, xx (2014), pp. 491–8.

40 Purdy, *After Nature*, pp. 266–82.

41 E. Benson, 'The Urbanization of the Eastern Grey Squirrel in the United States', *Journal of American History*, 100 (2013), pp. 691–710.

regimes and, indeed, the closer regulation of the human life in the city. Adelman's argument chimes with Chris Pearson's analysis of the management of stray dogs in nineteenth-century Paris.[42] Regulatory regimes did not just push animal-based industries towards the urban periphery, but also made it illegal to keep a pig or two in urban dwellings. Thinking through animal history raises interpretative questions about what constituted urban living in nineteenth-century Ireland and the interchange between rural and urban habits. Adelman's animal history also allows us to reimagine journeys through the city, transforming our sense of the sensual experience it offered. At the same time, the presence of animal life in the city often drew little comment, which makes Adelman's systematic scrutiny of the business directories of nineteenth-century Dublin another way of accessing how the experience of the city has been transformed since the nineteenth century.

A different form of mapping is prospected in Mary Orr's assessment of William Thompson's ichthyological studies. By focusing on Thompson's comparative, transnational work on Ireland's fishes, Orr challenges overtly territorialized and—implicitly or explicitly—nationalist approaches to natural history. Previous work on Thompson has failed to recognize how this Belfast naturalist was a recognized authority among his contemporaries, particularly the French scientists advancing the field of study; Thompson's research on Ireland's marine, estuarial, and freshwater habitats helped establish Europe's shared post-glacial biogeography, particularly with respect to the ranges of various fish species and the consequences of environmental change. Orr contends that Thompson's perspective was 'continental', which circumvented the condescension of British naturalists who tended to treat Ireland's natural history as a lesser, detachable knowledge domain, exposing their own parochialism. To historicize Ireland as a natural history knowledge domain, Orr argues, historians need to know 'who read, knew and worked with whom, and in several European language contexts'. Rather than regarding this as paradoxical, Orr's analysis suggests that the fluvial histories of Europe's biogeographies provide a means to rethink national, terrestrial categories. And therein lies the great challenge posed by the environmental humanities in the age of the Anthropocene. Just as Thompson recognized the sense of the connectedness of the world's life systems, so should this be at the heart of the humanist endeavour.

Thompson's methodologies were somewhat akin to what is now called 'citizen science' and among the themes examined in Patrick Maume's analysis of Emily Lawless's novels *Major Lawrence, F.L.S.* (1885) and *Grania: The Story of an Island* (1892) is how Lawless's fiction was informed by her collaborative

42 C. Pearson, 'Stray Dogs and the Making of Modern Paris', *Past & Present*, ccxxxiv (2017), pp. 137–72.

work as an amateur naturalist, her knowledge of geology, and her sense of the gradual professionalization of scientific practice—the transition, if you like, from Mr Darwin to Professor Huxley. For those unwilling to accept, or who remained ignorant of, the significance of this development, Lawless's fiction suggested that the amateur tradition could lead to a self-destructive monomania that threatened social isolation and produced redundant, outmoded work. Indeed, there are ways in which Lawless's fiction itself can be read as a response by an amateur to the professionalization of natural history, for it allowed her to deploy knowledge and understanding gleaned from her youthful natural history enthusiasms. In particular, *Lawrence* is punctuated with references to the Linnaean system of classification, directly placing Lawless's fiction in a lineage shared with Thompson and his French interlocutors; at the same time, a Darwinian pessimism overlays both novels. The peasant Irish, the subject of *Grania*, inhabit a world comprehended through biblical time and are seemingly rendered fatalistic by the natural forces that shape their lives. Whereas the peasantry knowingly allows its sheep to graze the bent an earlier improver planted to stabilize their island coastline, the narrator alludes to the geological foundations of their world and the deep time that has shaped it.

Maume suggests that the critical reception of Lawless's late work was strongly shaped by inter-generational tensions that placed her work on what, retrospectively, can be termed the cusp of the Irish revival. Although readings of Irish revivalism that see it as inherently anti-scientific and spiritual have been challenged, the Yeatsian desire to re-enchant the world is still placed at the heart of most general accounts of the revival—Yeats's 'Celtic note' continues to provide the dominant theme. However, as Seán Hewitt shows in his analysis of Seumas O'Sullivan's revivalist nature poetry, not only was this revivalist poet an experienced and active field naturalist but he drew on his scientific knowledge to frame his poetical re-enchantment of nature. A lepidopterist committed to Linnaean classification, O'Sullivan contributed to the *Irish Naturalist* and wrote poetry that foregrounded the physicality of the natural world, finding in the close study of its material form the means of its re-enchantment. As such, although O'Sullivan rejected the notion that the truths of the world were veiled, as the dominant revivalist trope had it, he did not seek to excavate the numinous from the natural world. Instead, he found spiritual meaning in nature's materiality, as manifest in his poetic animism and the sense of spiritual immanence he gleaned through close observation. In O'Sullivan's poetry, the physical and the transcendent could coexist, natural processes having an auto-sacralizing impetus, though, as Hewitt observes, the poet could at once satirize and defend his hallucinatory excesses.

Hewitt closes by observing that O'Sullivan was part of a younger generation of poets who sought to 're-materialize poetry' as they sought to 're-sacralize the material'. The growing concern with nature and the environment across the humanities reflects a more recent material turn driven by apparently quite different and more urgently felt motives. The vitality O'Sullivan tried to capture in his verse is something that we, as we develop new optics that allow us to see more clearly, cannot but recognize as fading. Whether we experience this as a form of spiritual loss, as surely O'Sullivan would have, or simply recognize the threat it poses to the future of human nature, the effect is perhaps the same. Something must be done. And, as historians, engaged in our relentless acts of recovery but sensitive to the concerns of the present, the sense of crisis means we cannot but start to see the past differently and ask of it different questions as we attempt, directly and indirectly, to figure out the causes of our current predicament. This need not be overtly environmentalist, although greater attentiveness to non-human nature as a form of life worthy of our consideration is likely to generate a greener consciousness. But, more than this, any sense that the life-giving capacities of the natural world are growing more fragile cannot but compel attention, having a ramifying effect on our consciousness that social justice and environmental sustainability are inseparable. Such an awareness should make us look afresh at nineteenth-century Ireland; our interest newly piqued.

# 1

# The Nature of Improvement in Ireland

*Helen O'Connell*

The history of 'nature' in early nineteenth-century Ireland is inextricable from that of 'improvement'.[1] In ways at odds with, or even opposed to, nature by seeking its transformation into society, Irish improvement was principally associated with the rationalizing of agricultural practice and the efficient management of natural resources. Improvers, those advocating and working towards cultivation and thus maximization of resources, were of course seeking to instil particular scientific approaches as widely as possible. But they were also striving to lessen Irish difference by improved means, working towards the eventual absorption of Ireland into the anglicized and homogenized mainstream. Improvement thus required the addressing of a range of social questions and moving beyond strictly material or environmental matters. Indeed, as a term that gradually came to connote an imminently practical approach to numerous social as well as infrastructural problems, 'improvement' was deployed in government reports, surveys, tracts, Dublin Society prize essays, novels, short fiction, and the numerous agricultural guides that were widely disseminated throughout the country. Hence, in the post-Union period, improvement became as much associated with social as with 'natural' issues. In fact, social questions were addressed as though they were essentially—or at least ought to be—entirely material and thus resolvable by means of the same general methods that were applied to drainage schemes or ploughing techniques. In what follows, it will be shown how Irish improvement writing, in fact the discourse as a whole, reflects the extent to which nature in this period was widely understood to be culturally determined.

1 The 'history of nature' might appear to be a contradiction in terms as nature by definition exists outside of history. However, the history of improvement in nineteenth-century Ireland is simultaneously the history of Irish nature, pointing to the ways in which, according to Timothy Morton, 'the final word of the history of nature is that *nature is history*'. See T. Morton, *Ecology without Nature: Rethinking Environmental Aesthetics* (Cambridge, MA, 2007), p. 21.

For improvers, their project could not proceed effectively until nature and society were properly distinguishable and kept distinct. But this vigilance did not of course preclude a countervailing tendency to have improvement in all its guises—landscape, language, housing, culture—entirely naturalized.

Early nineteenth-century Ireland was generally equated by improvers with an excess of naturalness, even brutishness, which was contrary to the more improved condition of Britain at a time of industrial and agricultural revolution. The inhabitants of the overwhelmingly unimproved Irish countryside were perceived to be as much in need of cultivation as the soil upon which they subsisted and from which they were insufficiently distinguishable. Irish writers, such as William Hickey (writing under the pseudonym, Martin Doyle), Robert Parker, Mary Leadbeater, and William Blacker, and societies, such as the Royal Agricultural Improvement Society, were committed to improvement and fully abreast of agricultural and manufacturing developments in Britain. They responded to the state of Irish nature by advocating all-encompassing schemes for progress that embraced people as well as soil, children as well as milch-cows, and kitchens as well as pigsties. The everyday life of the rural poor had to be made amenable to the practicality of improvement, responsive to forms of advice and instruction that were at odds with custom and traditional practices. The achievement of an Irish improvement of drained bogs, tidy cottages, and enclosed fields would signify not simply the country's modernized state but its naturalized absorption within the terms of the United Kingdom of Britain and Ireland, as indeed was presumed to have occurred with Scotland in the aftermath of Union in 1707.

Improvement was historically aligned with forces of colonization, anglicization, and, after 1800, Union.[2] While primarily Protestant rather than Catholic, and clearly committed to particular types of progressive projects (broadly understood), Irish improvement was not by any means entirely coherent or institutionally organized.[3] Clearly stimulated by Arthur Young's *Tour in Ireland* and his implicit faith in the unifying potential of improvement, Irish writers duly worked towards achieving an 'organic society' by means of ploughs, practical advice, and continual observation of material and social life.[4] The term, 'improvement', was perhaps overly flexible, capable of signifying a variety of activities and projects from crop rotation, bog drainage, vaccination

2 For an account of improvement before 1800, see T. Barnard, *Improving Ireland? Projectors, Prophets and Profiteers, 1641–1786* (Dublin, 2008) and J. Livesey, *Empire and Civil Society: Ireland and Scotland in the Eighteenth-Century Atlantic World* (New Haven, CT, 2009), pp. 54–89 and 176–80.

3 Despite its strong Protestant associations, there was some improvement activity amongst Catholic landlords in the eighteenth century, such as Charles O'Conor of Belanagare. See Barnard's account in *Improving Ireland?*, pp. 183–4.

4 A. Young, *A Tour in Ireland* (2 vols., Dublin, 1780).

schemes, the modernization of ports, and—most ambitiously—the inspiring of an avowedly practical approach to life amongst the rural poor. But many of these diverse projects were united by a shared vocabulary and set of underlying assumptions, which were articulated with remarkable consistency from the early nineteenth century through to the Famine and beyond. Improvement depicted itself as a wholly neutral and progressive force in an otherwise divided society. Indeed, in the works of Leadbeater, Blacker, and Parker, amongst others, improvement attempts to constitute itself as a civic space in which differences and associated animosities could be in part resolved. Such resolution was of course deemed necessary for the pursuit of an economic and social progress that was itself supposed to be disinterested. A seemingly innocuous term, which was supposedly without explicit sectarian or party-political associations, improvement was intended to signify the possibility of a society capable of developing beyond the influence of political, sectarian, and historical difficulties. In its seemingly rigorous material form, improvement was represented as a solution to those historical and persistent tensions, which were equated with Irish 'nature'.

There was a particular, if implicit, motivation for much of the improvement activity in the post-Union period and one that was to torment the possibility of its success. Improvement in these years was not only stimulated by the prospect of economic and social progress and the achievement of stability. The many tracts, pamphlets, and articles written in the aftermath of the Act of Union were seeking to redress more than a perceived rural backwardness and uncultivated nature. In fact, improvement at this point was as much a response to an already modernizing culture as to a pre-modern one and thereby differed in key respects from the seventeenth- and eighteenth-century activities of William Petty and James Caldwell respectively.[5] As such, improvement was itself shifting ground from an earlier model of 'civil society' within the terms of empire—as identified by James Livesey—to a distinctly Unionist (civil) society in the broader context of the campaign for Catholic Emancipation.[6] The campaign for Catholic Emancipation was of course itself a process in which a form of Irish nationalism was being articulated and defined. In emphasizing the need for a strengthened countryside in which traditional class structures and the form of authority thereby implied largely endured, improvers were arguably seeking to counteract the development of Catholic nationalism. The explicitly secular ethos of improvement is rhetorically opposed to the perceived emergence of an assertively Catholic middle class of farmers, shopkeepers, and traders. The broader reality of early

5 For a detailed account of Petty and Caldwell, see Barnard, *Improving Ireland?*, pp. 41–72.

6 For a discussion of that earlier eighteenth-century mode of improvement as an articulation of an emerging 'civil society', see Livesey, *Empire and Civil Society*, pp. 84–9.

nineteenth-century Catholicism, however, shapes improvement discourse in several ways, at one level Catholicizing and democratizing it in a manner that disconcerted improvers themselves.

In order to disseminate improved methods, improvers had to communicate with the numerous Catholic farmers and labourers of the countryside. This process was to have a decisive effect on the fate of improvement, making for any number of contradictions and difficulties. Leadbeater's *Cottage Dialogues* (1811) and Martin Doyle's *Irish Cottagers* (1833), for example, contradictorily depict a society shaped by both the most progressive of agricultural practices and a strongly traditional way of life in which only home-grown vegetables and cups of milk fill the stomachs of labouring men and their families. Despite the widely perceived threat to traditional rural society from the efficiency of improved methods, improvers frequently couched their recommendations in organic metaphors of wholeness, unity, and community, of fulfilling reciprocal exchanges between landlord and tenant and the pleasing satisfactions of improved labour over the inefficiencies and fragmentariness of traditional working methods.[7] The improved Irish farmer would be as much defined by his wariness of modern consumption as his rigorous application of the most up-to-date of sewing techniques. While insisting that any number of traditional techniques and methods be jettisoned, improvement was nonetheless keen to depict itself as strongly traditional and as conserving some immemorial morality or even 'moral economy'.[8] In fact, improvers flagrantly invented an idea of tradition in which both religious and political identities, and all the historical weight they contained, were supposedly rendered unthreatening by the achievement of a particular level of economic and domestic stability that was to be consistently underpinned by the grounding force of practicality.

Uncomfortable with both cosmopolitanism and nationalism, improvers worked to emphasize the value of localness and ruralness, which was as much a means of signifying a specific sensibility as an everyday actuality. Migration to the city is steadily cautioned against by Doyle and Leadbeater, for example, and poor families are advised to seal themselves domestically against the disrupting pressures and sensations of public space—in fact community—in a necessarily modernizing and individualizing environment. The home, in

7 Improvement was understood by such contemporary observers as Oliver Goldsmith and John Clare (amongst others) to pose a threat to traditional rural life. For an account of anti-improvement themes in English literature of the period, see J. Barrell, *The Idea of Landscape and the Sense of Place, 1730–1840* (Cambridge, 1972).

8 For the earlier articulation of the notion of a moral economy, see E. P. Thompson, 'The Moral Economy of the English Crowd in the Eighteenth Century', *Past & Present*, l (1971), pp. 78–9. For an application of the idea of a 'moral economy' in the eighteenth-century Irish context, see T. Bartlett, 'An End to Moral Economy: The Irish Militia Disturbances of 1793', *Past & Present*, xcix (1983), pp. 41–64.

all its material and symbolic manifestations, was understood to be central to regulating and thwarting any disruptive forces. In *Hints to the Small Farmers of Ireland* (1828), Doyle comments that all those who belong 'to a working farmer's family' should dress according to their 'station, and consistently with your circumstances'. Rural workers (and their families) should never emulate 'the appearance of mantua-makers and shop-boys, a far less valuable class', he declares, than that of agricultural labourers.[9] Shop boys and dressmakers, many of whom would be immersed in trade and commercial society, were threatening the stabilizing class hierarchies of rural life. Doyle advises that such a threat had to be counteracted by means of the wearing of clothes of '*home*-made manufacture, as much as possible'. While conceding that such clothes 'might not be so fine as bought articles', Doyle insists that they are more creditable, more durable, and more economical'.[10] Home-made clothing holds symbolic power far in excess of its material value. The wearing of such clothing enforces the creditableness of the countryside over the dishonourableness of the city and the durability of agriculture over the instabilities of trade. Indeed, the home-made, grown, and cultivated were morally elevated over imported foodstuffs and factory-made commodities. But this is not of course an argument for nature over culture, or for the value of inhabiting a state of nature over that of commercial society. It is instead an insistence that the countryside paradoxically provides better protection against the forces of nature than the city ever could. In fact, for Doyle, as well as others, the city is more closely associated with an uncultivated nature than the countryside, urban life oddly perpetuating a nature that can only be effectively disciplined or cultivated by means of agricultural and social improvement.

From the outset in seventeenth-century England, improvement as a term connoted the maximizing of gain on cultivated ground and processes for the cultivation of 'waste' or infertile land.[11] Paul Warde has described how improvement was associated with efforts 'to transform the value of marginal lands' into usefulness and profit.[12] Improvement specifically entailed progressive agricultural, horticultural, and manufacturing methods, which were rooted in the pioneering scientific experimentation of seventeenth-century England and which continued to evolve in line with developments in chemistry and manufacturing.[13] Improvement defines itself as entirely practical rather than theoretical,

9 Doyle, *Hints to the Small Farmers of Ireland* (Dublin, 1828), pp. 22–3.

10 Ibid., p. 23.

11 An argument for the particular Englishness and non-translatability of improvement is posited by Slack. See P. Slack, *The Invention of Improvement: Information and Material Progress in Seventeenth-Century England* (Oxford, 2015).

12 P. Warde, 'The Idea of Improvement', in R. Hoyle, ed., *Custom, Improvement and the Landscape in Early Modern Britain* (Farnham, 2011).

13 Warde describes how the period from 1640 to 1660 was a 'revolutionary moment' for

a body of knowledge acquired through continual application and re-application as well as hands-on engagement with specific materials and real conditions. By empirical means, improved agriculture was supposed to be wholly distinct from customary practices, which were largely deemed to be outmoded and inefficient. In fact, in Ireland, such methods were widely equated with Catholicism and were thus symbolic of a deeply rooted and unhelpful loyalty to custom and precedent. Irish improvers sought to displace traditional forms of agricultural knowledge by instilling a scientific understanding of soils, seeds, and tillage. For example, Doyle worked hard to root out the unprofitable customary methods passed down the generational line ('predecessors') from grandfathers through to fathers and sons in favour of scientific and non-customary techniques.[14] The displacement of traditional agriculture was supposed to stimulate and go hand-in-hand with the displacing of traditional Catholic culture more generally, which was understood to be worryingly naturalized.

Paul Slack has shown that while improvement originated in particular agrarian innovations of the seventeenth century, it gradually encompassed manufacturing, mining, and town-planning, in part stimulating the industrial revolution of the later eighteenth century. For Slack, the history of English improvement 'is by definition a Whiggish story of progress', constituting 'a mode of discourse which looked for inspiration to a better future and not to an ideal past'.[15] The Whiggishness of improvement, itself partly rooted in both the reformation and enlightenment, was, however, to encounter obstacles to such optimism in Ireland, resulting in ambivalent or even outright anti-Whig sentiment in the work of some Irish improvers.[16] Indeed, many nineteenth-century Irish writers, who were as stimulated by the progressiveness of improvement as Samuel Hartlib or William Petty, and true believers in the great economic and social gains to be achieved by crop rotation and the cultivation of turnips, nonetheless found themselves despairing at points over the difficulty of realizing 'a better future' in the particular Protestant terms of improvement. In part this might explain why Irish improvers were more keen to invoke notions of tradition, and to place restraints on consumption and urbanization, than their English peers.

Nonetheless, the general Whiggishness of improvement provided the rhetorical impetus for many of these writers, even if there was not always

improvement in the 'republican circles associated with Samuel Hartlib'. Warde, 'The Idea of Improvement', p. 128.

14 Doyle, *Hints to the Small Farmers of Ireland*, p. 36.

15 Slack, *The Invention of Improvement*, pp. 263, 264.

16 An anti-Whig Irish improver is the Revd William Richardson, whose work was, at least in part, stimulated by his opposition to the Whigs. For a thorough account of Richardson, see A. Blackstock, *Science, Politics and Society in Early Nineteenth-Century Ireland: The Reverend William Richardson* (Manchester, 2013).

certainty on the possibility of achieving 'a better future' in a divided society. Many improvement writers invoke a seemingly liberal, even democratizing, rhetoric that admirably shuns oppression of the poor, women, children, and animals. Several improvers express particular disdain for anti-Catholicism, equating crude sectarianism with backwardness and impasse. The starting point and rhetorical mode for most Irish improvers is characteristically optimistic; in practical terms, optimism is essential to communicating the value of modern ploughing techniques and stone walls as well as to conveying the common sense of temperance and savings. Indeed, the writing of a pessimistic improving pamphlet or guide is unthinkable. Optimism, and particularly the endurance it supposedly instils, is obviously fundamental to any improvement venture. In Leadbeater's tract, *Cottage Dialogues*, it is advised that 'when any thing fails with you, strive to make out the reason why it went astray, and do better next time'.[17] There is no giving up, or resignation to the possible futility of irredeemably hard and endless work. The rhetorical terms of improvement simply do not permit existential despair, anguish, or anxiety, which does not ensure that pessimism or uncertainty never intrude between the lines and against the grain of the texts themselves. But there was more to this; it was essential to improvers that faith be maintained in the claim that following particular instructions would ultimately yield specific results, that outcomes could be predicted and anticipated on the basis of certain conditions and applied techniques. In a wider context, the working hypothesis was the capability to transform nature (and, by extension, society) by material means and processes.

In Ireland, improvement was to remain primarily associated with the countryside, its largely rural focus pointing to the fact that in the early nineteenth century, as Brinley Thomas has explained, 'Irish land was being called in to alleviate the mounting pressure on English land'.[18] Ronald Findlay and Kevin O'Rourke claim that as late as the 1830s imports from Ireland 'were equivalent to 13% of English agricultural output, and more than 85% of total English imports of grains, meat, butter, and livestock'.[19] Allan Blackstock has explained that the 'integration of Irish agriculture into the United Kingdom's domestic food supply was considered a consolidating benefit of the Union'.[20] Nineteenth-century Irish improvement can be understood as committed to the consolidation of the Act of Union in the most complete sense. In fact, Irish

17 Mary Leadbeater, *Cottage Dialogues among the Irish Peasantry, Part 2* (Dublin, 1813), p. 23.

18 B. Thomas, 'Escaping from Constraints: The Industrial Revolution in a Malthusian Context', *Journal of Inter-Disciplinary History*, xv (1985), pp. 741–2.

19 R. Findlay and K. O'Rourke, *Power and Plenty: Trade, War, and the World Economy in the Second Millennium* (Princeton, NJ, 2009), p. 238.

20 Blackstock, *Science, Politics and* Society, p. 120.

agricultural products were necessary for England to achieve the very kind of Whig progress it desired and in part at least believed itself to be fulfilling. Irish land, and the grains, butter, and meat it produced, nourished England's Whig vision of itself while arguably emphasizing the lack of—and indeed curtailing of—such progress, vision, and associated nourishment in Ireland itself.

English improvement historically defined itself against a presumed naturalness in the unshakeable faith that nature was ultimately governable and disciplinable into order and profit. But for Irish improvers such faith was harder won and more difficult to sustain. Ireland's perceived excess of nature, which supposedly shaped language, religion, politics, and the conducting of routine everyday tasks, meant that society in its most practical of manifestations was taken to be undeveloped. Far from welcoming the Romantic prospect of an abundant naturalness and related absence of conventionality, this excess of nature was understood by improvers as the regrettable cause and effect of the generally non-cultivated and Catholic condition of the country. Improvement positioned itself as a 'social' discourse confronting a nature that was, paradoxically, culturally entrenched.

As in England and Scotland, improvement in Ireland stimulated the founding of societies and associations that were committed to the instilling of improved methods in particular realms. Almost every perceived agricultural, manufacturing, and social problem could be addressed by means of 'improvement' with a society established to tackle it. Hence, in addition to the very generalized scope of the Dublin Society and the Society for the Improvement of Ireland, there was the Royal Agricultural Improvement Society of Ireland, the Society for the Improvement of Prison Discipline, the Dublin Society for the Improvement of Husbandry and Internal Resources, the Irish Waste Land Improvement Society, and the Royal Society for the Improvement of the Growth of Flax in Ireland. The existence and proliferation of such societies suggests the ways in which the term improvement became attachable to a variety of reforming and development needs. As a general frame of reference, improvement provided the terms for the highlighting and consequent addressing of a range of problems; however, those other needs or difficulties, which were not somehow answerable in the terms of improvement, were then rendered obscure or non-existent.

In addition to those publications that address very particular topics or problems, such as John Hamilton's *An Essay on the Subservience of Improved Agriculture to the Proper Feeding of Stock* of 1800 and William Richardson's *An Essay on the Improvement of the Great Flow Bogs of Ireland* of 1803, are those that outline methods for the improving of Irish humanity. Examples would be Robert Bellew's *Thoughts and Suggestions on Means Apparently Necessary to be Adopted by the Legislature, Towards Improving the Condition of the Irish Peasantry* of 1808, Robert Parker's *A Plan for the General Improvement of the*

*State of the Poor of Ireland* in 1816, and Doyle's *The State of the Poor of Ireland Briefly Considered and Agricultural Education Recommended, to Check Redundant Population and to Promote National Improvement* (1820). For improvers, those techniques for the cultivation and drainage of the soil and bogs of Ireland needed also to be deployed in the homes and minds of the poor. In Leadbeater's *The Pedlars* of 1826, it is stated of an exemplary, as improved, small farmer that 'his mind had been previously cultivated for receiving the seeds of learning'.[21] It would seem that improvers had to think of the rural poor in such 'natural', even agricultural, terms in order to envisage the possibility of their progress. The designating of the rural poor as a category much like bogs, flax, and soil made possible the articulation of a programme for their improvement. At one level it could be claimed that such a mode of analysis demonstrates that improvers commendably attempted to assess social problems in the most material and non-ideological of terms. However, it also foregrounded a prevailing assumption that the rural poor were essentially natural or semi-natural, effectively inhabiting barbaric or raw conditions. But there is also unambiguous economic thinking at work: this naturalizing and attempted objectifying of the rural poor points to the way in which the numerous inhabitants of the Irish countryside came to constitute a resource and category, that of labour. This category then needed to be grasped in seemingly objective and economic terms. As a result, improvement in these years notably shifts part of its focus to the instilling of those conditions that will allow for the fulfilling of this very labouring and thus commercial potential.

For Irish improvers, any number of large-scale improvement projects, such as the drainage of the Bog of Allen, would be futile if the rural poor were left in their unventilated mud huts.[22] As late as 1841, the *First Report of the Royal Agricultural Improvement Society of Ireland* declares that as a body it is

> firmly persuaded that no measures can be adopted for permanently and effectually promoting the agricultural interests in Ireland, which do not tend to advance the moral and social condition of the labouring population, and to elevate them in society, so that by improving their habits by example and encouragement, increasing their comforts by sympathy and attention, they may learn to feel that the interests of all classes are identified, and making Agriculture, as it were, a neutral ground, to merge all differences in the common good.[23]

21 M. Leadbeater, *The Pedlars* (Dublin, 1825), p. 13.

22 Itself a controversial project, drainage of the Bog of Allen was a preoccupation in the early nineteenth century. For an account of the Revd William Richardson's anxiety over this particular scheme, see Blackstock, *Science, Politics and Society*.

23 *First Report of the Royal Agriculture Improvement Society of Ireland* (Dublin, 1841), p. 14.

Improvement would fail as a project if it neglected to encompass 'the moral and social condition of the labouring population'. The performance of labour could not be separated off from morality, which was inextricable from the domestic conditions inhabited by the rural poor. It was unthinkable that an immoral and dirty labourer could work to the level necessary for improvement to occur and be actualized. The labouring classes needed to be, in a sense, cultivated by 'example and encouragement' to the extent that they would be socially elevated from their natural and lowly condition. For improvers, agriculture provided the very kind of 'neutral ground' in which improvement could be materially and securely instilled, free of the impediments of religious or linguistic divisions, while working towards the kind of homogenization that would nonetheless be supported by critical distinctions of class.

Few agricultural improvers restricted themselves wholly to agricultural advice; in fact, most found themselves venturing into domestic and even intimate realms. In Doyle's tracts and articles, written between the late 1820s and 1850s, 'hints' for the mass instilling of rationalized agricultural methods go hand-in-hand with the imparting of advice on domestic hygiene and personal appearance, stepping well over the threshold of the cottage or cabin door to confront the rural poor on some of the most intimate details of their daily lives. In *Hints to the Small Farmers*, a chapter devoted to 'Cottage Cleanliness' and 'Personal Tidiness' appears alongside others on such evidently material and seemingly neutral topics as 'Condition and Quality of Land' and 'Preparation of the Soil'. For Doyle, the civility of the rural population had to be tackled by addressing their supposedly squalid appearance as well as the deplorably unhygienic conditions they inhabited. He rails against dunghills at the front door, which give rural houses 'a dirty and disgusting appearance', and the 'everyday filthiness' of too many poor women, men and their children.[24] Despite the difficulties of translating recommendations for self-improvement into the same kind of instructional mode deployed for soil preparation, Doyle nonetheless attempts to approach such matters in the very same matter-of-fact manner. William Blacker's *An Essay on the Improvement to be Made in the Cultivation of Small Farms* (1834) is in the same vein: it includes recommendations on the 'advantages of cleanliness and ventilation' of the home alongside suggestions for the planting of green crops, the care of milch-cows, and the superiority of turnips over potatoes as feed for farm animals.[25]

24 Doyle, *Hints to the Small Farmers of Ireland*, p. 18.

25 William Blacker, *An Essay on the Improvement to be Made in the Cultivation of Small Farms by the Introduction of Green Crops and House-feeding the Stock Thereon* (Dublin, 1834), p. 45.

The starting point of improvement certainly had to be material and bodily, but they were also its endpoint, palpably connoting the moral progress and elevation from nature they were simultaneously making possible. In such a scheme of things, bodiliness, for example, did not entail proximity or continuity with nature, but the most effective disciplining of it. Likewise, it would not simply be that civilization would follow on from inhabitation of stone houses, the use of a table, the wearing of clean clothes, and the eating of wholesome food. The achievement of such conditions would itself constitute civilization, which, in this scheme of things, is almost wholly a practical matter. Indeed, improvers largely seek to halt progress at the point at which practicality has been achieved, fearing greatly what might lie beyond that particular point of development.[26]

Indeed, practicality is both the means and end of much improvement discourse, which largely directs its attention to the utterly particular and specific, such as soil type, planting regimes, and the hygienic management of dairies. Many of these entirely material issues would appear to possess little, if any, broader social or political significance for such contemporary and 'hot' issues as Catholic Emancipation. But the rhetorical modesty and reticence of improvement conceals the actual largeness of its ambition, and the extent to which it entailed a thorough transformation of the inhabitants and landscape of the Irish countryside. By means of improvement, Ireland was to be transformed from an overtly Catholic and uncultivated (or 'natural') country to a largely reformed (or 'social') constituent of the Union. For improvement writers, the social and political condition of a country had everything to do with the state of the soil; political stability an impossibility in an uncultivated and raw environment.

Some improvement associations were committed to enormous and ambitious projects that would result in a significant alteration of the Irish landscape. An example would be the Irish Waste Land Society, which embarked on large-scale drainage projects on the Gleneask estate in the early 1840s. However, improvement in the early nineteenth century is also notable for the extent of its preoccupation with very small projects and for the related manner in which it is addressed to the 'lower orders' and women. By means of the writing and printing of cheap and readily reproducible pamphlets, 'improvement' was supposed to be efficiently and comprehensively disseminated to the numerous small farming families that populated the countryside. As a textual phenomenon, improvement was prominent in the public sphere of print in newspapers, periodicals, and pamphlets

26 In this way, 'improvement' both enforces and contradicts aspects of the history of the progress of western society towards civilization, as documented by Norbert Elias. See Elias, *The Civilising Process*, trans, Edmund Jephcott (Oxford, 2000).

against the broader backdrop of the post-Union collapse of Irish publishing.[27] Numerous improvement pamphlets were printed in Dublin and Cork and other provincial towns in the first three decades of the nineteenth century at a time when almost all Irish novels, for example, were published in London.[28] These publishing ventures are notable for the fact that they assumed the existence of a reading public amongst the rural lower orders and for the various means they deployed to address it. The assumption was effectively that a civil society already existed, which was of course literate and needed to be addressed, as well as managed in some respect, by textual means.

Improvement equates itself from the outset with both literacy and print, envisaging its fortunes as wholly bound up with the wide dissemination of printed materials. The packaging of agricultural improvement for the generally poor and small farmers of Ireland was also borne out of the particular conditions of the post-Union period. Within a context of both widespread landlord absenteeism and demographic expansion, improvers felt compelled to address the ever increasing numbers of small farming families. This targeted audience was necessarily poor, primarily Catholic, and existing on the most meagre of resources in the heavily populated Irish countryside. Leadbeater's fictional tracts, such as *Cottage Dialogues*, *The Landlord's Friend*, and *The Pedlars*, depict many of these very poor yet improving small farmers, women, children, and their domestic animals within a context of perceived backwardness, scarcity, and recalcitrance. Intended for free distribution by its subscribers to the rural poor and schools throughout the country, *Cottage Dialogues* was the first improvement pamphlet to address this audience directly, urging upon them the stability as well as evident common sense of improvement over the perceived other-worldliness of traditional Catholicism. In a preface to an 1813 edition of the tract, Leadbeater thanks her subscribers for 'taking a large number of copies for distribution' and thereby promoting its 'circulation among that class of readers, for which they were intended, in a more expeditious manner than ... could have been effected by other means'.[29] This method of expeditious circulation and distribution was essential as Leadbeater's intended readers were not those few whose names appear on the list of subscribers—most of whom were themselves landowners—but the

27 Both Parker and Hickey began their literary careers as improvers by writing columns in newspapers. Parker's pamphlet originated in a series of letters to the Editor of the *Cork Advertiser*; Doyle wrote a series of articles on agricultural improvement for small farmers for the *Wexford Herald*, which was later reprinted in pamphlet form as *Hints to the Small Farmers of the County of Wexford*, with a subtitle stating that it was suited to 'the circumstances of most parts of Ireland'.

28 N. Ó Ciosáin, *Print and Popular Culture in Ireland, 1750–1850* (Basingstoke, 1997).

29 M. Leadbeater, Advertisement, *Cottage Dialogues, Part 1* (4th edn., Dublin, 1813), p. vii.

numerous literate farmers who largely existed beyond the conventional terms of commercial society.[30]

In his Preface to the first edition of *Hints to the Small Farmers*, Doyle states that 'our various essays upon agriculture are often more suited to the taste of the gentleman farmer, than to the wants of the practical husbandman; they are too expensive for the purse, too learned for the unlettered mind of the humble peasant'.[31] Doyle's *Hints* is supposed to redress that tradition of agricultural writing by being imminently practical, affordable, and, crucially, less lettered than many agricultural guides. Doyle's *Hints* will be able to address the very many small farmers rather than the very few gentlemanly ones by being fully attuned to mass, and not elite, taste. The very terms of Doyle's project thus call attention to the ways in which Irish improvement became centrally preoccupied by the need to communicate with the many poor people (including women and children) of the countryside, rather than the few Protestant landowning men. Some agricultural societies clearly agreed with Doyle's assessment and viewed the circulation of his manuals to be an effective means of spreading, and thereby entrenching, improved techniques. In April 1830, for example, it was reported in the *Drogheda Journal, or Meath and Louth Advertiser* that the Louth Agricultural Society 'resolved that 300 copies of Martin Doyle's Hints to Small Farmers should be distributed within the district of the society'.[32] In a more general vein, the *First Report of the Royal Agricultural Improvement Society of Ireland* stated as one of its objectives, 'the dissemination of practical and useful knowledge, through the medium of cheap publications ... The improvements in husbandry, both practical and scientific, of late, are so numerous, and so beneficial in their effects, that it is most desirable to have them circulated through the agricultural population as extensively as possible'.[33]

In shifting from the few to the many, and fully encompassing women and even young children, a shift was also occurring from Protestant to Catholic and rich to poor, reflecting a perceived and unstoppable development towards democratization. However ambivalently any of these improvers felt about such democratizing developments in Ireland, their work was by its very nature partaking of such developing structures and thereby enforcing them. In ways, publications such as *Cottage Dialogues* and *Hints to the Small Farmers* signify

30 The cheap didactic pamphlets produced by the non-denominational Kildare Place Society were also widely disseminated on similar principles and used as reading texts in schools for the very poor. For accounts of the Kildare Place Society, see H. J. Hislop, 'The Kildare Place Society 1811–1831: An Irish Experiment in Poplar Education' (University of Dublin Ph.D. thesis, 1990) and Ó Ciosáin, *Print and Popular Culture in Ireland*.

31 Doyle, Preface, *Hints to the Small Farmers of Ireland*, p. iv.

32 *Drogheda Journal, or Meath and Louth Advertiser*, 17 April 1830.

33 *First Report of the Royal Agricultural Improvement Society of Ireland* (Dublin, 1841).

an attempt to exert some measure of control over those processes, particularly as they were perceived to shape the conditions and, perhaps most of all, dispositions of the Catholic majority of Ireland.

The need to encompass the labouring classes and lower orders generally within improvement had long been understood to be essential to its fortunes. But the rural poor of Ireland had not previously been addressed in the precise manner of either a Doyle or a Leadbeater. As noted, William Hickey deliberately adopted the pseudonym of 'Martin Doyle' in order to distance his writings from his identity as a clergyman and to pass himself off as 'ordinary'—socially closer to the many Catholics than the few Protestants. The poor and labouring classes were now being explicitly embraced within improvement as though they possessed agency, could determine their own circumstances, and, indeed, that of improvement as a project. In *Hints to the Small Farmers*, Doyle claims that 'if the soiling system in summer, with green crop feeding in winter, were in general use among you, in a climate so favourable, from its mildness and moisture, as ours is, to herbage crops, and turnips, and mangel wurzel, and so frequently unfavourable to wheat and barley, I should have no doubt of your rapidly rising in the scale of comfort'.[34] The adoption of a practical and entirely straightforward system should thus ensure elevation to bourgeois comfort. Significantly, Doyle is suggesting that the achievement of a respectable form of 'middling' comfort was well within the reach of the humblest in society. The achievement of these middling conditions merely required that agency be directed to the work of improvement rather than participation in local secret societies, for example. William Blacker echoes this very view when he declares that 'no practicable method exists for ameliorating the condition' of the labouring poor 'except by inducing them to become the active instruments of their own amelioration'.[35] This perceived agency of the poor needed to be directed, absorbed so to speak, into the bigger project of improvement, aligned with one of its prevailing pre-Darwinian assumptions, that of the agency of humanity as a whole to determine and control its social and, indeed, natural, environment. If the significant rural population, who had demonstrated only too frequently the ways in which they could act on the basis of their own assumptions, feelings, and occasionally overly political ideas about life, were not brought within an improving framework and terms of reference, there was little prospect of maintaining the existing social structure of post-Union Ireland.

34 Doyle, *Hints to the Small Farmers of Ireland*, pp. 41–2.

35 William Blacker, *An Essay on the Best Mode of Improving the Condition of the Labouring Classes of Ireland* (London, 1846), p. 10. Blacker's essay was originally submitted to an essay prize of the Royal Agricultural Improvement Society of Ireland and was awarded the Gold Prize. It was afterwards printed in England. See ibid., Preface.

It was no longer deemed sufficient to write for the landowner or gentleman farmer in the hope that they could then radiate humanizing and civilizing values across their estates and large farms. While the Union was envisaged as potentially nurturing ideal conditions for improvement, providing opportunities for the replication of the eventual prosperity of post-Union Scottish conditions, many landowners abandoned their Irish estates to the management of agents. From a classic improvement perspective, this would have been a significant setback: the resident landowner, who favoured active life on his country estate over an inactive life in the city, was as central to the more ideal strains of Irish improvement as to its English or Scottish equivalents.[36] A highly knowledgeable agriculturist, the improving and largely resident landowner was to be always busily devising and directing various agricultural and horticultural projects on his estate. He would of course possess comprehensive first-hand knowledge of his land and tenants and be constantly learning about the latest agricultural techniques and methods from pamphlets, magazines, and perhaps attendance at meetings of local agricultural societies. Such figures, like, say, Richard Lovell Edgeworth on his Edgeworthstown estate, might even be amateur scientists and inventors themselves, contributing to developments in agricultural science, while all the time actively applying the most scientific of methods in their day-to-day agricultural practices.[37] This landowner would be invigorated, energized, and stimulated by life and work on his estate, relishing the role of dutiful, benevolent cultivator. But that ideal was necessarily harder to realize in the absentee conditions and altering social structures of early nineteenth-century Ireland. Many improvers certainly tried to encourage landowners to return to their Irish estates. In *A Plan for the General Improvement of the State of the Poor of Ireland*, William Parker suggested the levying of a tax on absentees as 'the surest mode of procuring the benefit to society, from a residential gentry'. For Parker, it was 'a residential gentry' which could 'effectually promote the prosperity and civilisation of [Ireland]'.[38] The improvement of the condition of the 'poor of Ireland' was, according to Parker, unthinkable without the active involvement of resident landowners. In *The Landlord's Friend* of 1814, written

36 For a discussion of some improver landowners in eighteenth-century Ireland, see Barnard, *Improving Ireland*; and for accounts of their peers in England and Scotland, see E. Griffiths, '"A Country Life": Sir Hamon Le Strange of Hunstanton in Norfolk, 1583–1654'; B. McDonagh, '"All Towards the Improvement of the Estate": Mrs. Elizabeth Prowse (Northamptonshire), 1764–1810'; and 'Improvement on the Grand Estates in Strathspey in the Later Eighteenth Century: Theory, Practice and Failure?', in Hoyle, *Custom, Improvement and the Landscape in Early Modern Britain*.

37 For an account of Richard Lovell Edgeworth as a landlord who made the transition from neglectful absenteeism to responsible benevolence, see M. Butler, *Maria Edgeworth: A Literary Biography* (Oxford, 1972), p. 77.

38 W. Parker, *A Plan for the General Improvement of the State of the Poor of Ireland* (Cork, 1816), p. 59.

explicitly to discourage absenteeism and generate optimism regarding the social and political possibilities of post-Union Ireland, Leadbeater attempted to show how a resident landowner would experience no discomfort or boredom in rural Ireland. In fact, Leadbeater explains that active and resident landlords, supported by their compassionate and hard-working wives, would quickly discover the social and existential pleasures of diffusing improvement throughout their estates. But, despite such efforts, post-Union Ireland called for a different structure for improvement, one that needed to emanate much more from both the middle and bottom than from the top and to which the work of women was fundamental. Irish improvement thus becomes more distinctly middling rather than elite. The outbreak of Rebellions in 1798 and 1803 suggested all too clearly how inadequate and ineffective earlier patriarchal notions of the morally upstanding landlord had been. Despite her plea for an end to absenteeism in *The Landlord's Friend*, Leadbeater had herself been articulating a model of improvement in which middling women loom large and where landowning families were either absent or exemplary of a generally regressive way of life.

The communication of improvement needed to be pervasive and thorough: indeed, improvers had to proceed with their project as though differences of language, class, gender, and religion could no longer constitute a significant barrier to the communication and reception of reform. In a sense, the mere fact of addressing the poor, and depicting a process whereby improvement was being achieved in the realistic conditions of the countryside, seemed to suggest the improvability of Ireland as a whole for the writers themselves. The process of the representation of such a society was as—if not more—significant as its subsequent reception. Indeed, the ambition was no less than that improvement should fully saturate the countryside, completely convincing all of its correctness as well as viability.

Nineteenth-century Irish improvers were of course as keen as their seventeenth- and eighteenth-century forebears on large-scale improvement projects for the reclamation of waste land and the drainage of bogs. But much of their work is particularly notable for the extent to which it focuses on the very small and sometimes unpromising patches of ground that surrounded the cabins and small cottages of the countryside. Some tracts include drawings of cottages and accompanying farm buildings, neat symmetrical structures built of stone, with chimneys and windows, which serve as a contrast to the windowless, chimneyless cabins that were proliferating throughout the countryside.[39] Instead of focusing their attention on the hundreds or

39 The pamphlets of both Leadbeater and Doyle contain such drawings; Doyle's *Hints to the Small Farmers* includes a particularly detailed architectural drawing of an exemplary cottage and accompanying farm buildings.

thousands of acres in the possession of landlords and gentlemen farmers—and the significant capital thereby signified—these improvers were more anxious to develop schemes and models for maximizing the full potential of half-acre and acre plots and small farms, many of which were widely cultivated for the hand-to-mouth consumption of potatoes by some of the very poorest. Leadbeater provided detailed advice on manure, digging, potato cultivation, and the rearing of pigs for poor farmers compelled to subsist on such small plots. Almost all of Doyle's publications are preoccupied by the condition of the many small farms throughout the country and are committed to the envisaging of improved agriculture on a much reduced scale. Blacker's *An Essay on the Improvement to be Made in the Cultivation of Small Farms by the Introduction of Green Crops and Housefeeding the Stock Thereon* is also stimulated by his commitment to improving the agricultural practices of small farmers. Blacker, who worked as an agent on the large estates of Lord O'Neill and the Earl of Gosford, had the bailiffs on these estates educate the tenants on improved agricultural methods. He also arranged for the hiring of sixty Scottish agriculturists for whom he procured positions on various estates throughout Ireland as a means of widely disseminating improved methods.[40] The work of Hickey, Leadbeater, and Blacker exemplifies the manner in which improvement was necessarily compelled to work on this more diminished level. This diminishment was not a curtailment of the ambition of improvement as a project but a response to the particular and 'natural' pressures of post-Union life.

The post-Union countryside was continually filling up with cabins and accompanying plots, the consequence of the ever increasing population of the late eighteenth and early nineteenth centuries, itself reflective of a seemingly uncontrolled nature.[41] Indeed, Irish nature in this period is heavily populated, the site of a demographic explosion that was nurtured by the widespread cultivation of potato plots and disregard of one of the prevailing social conventions of modern Britain—that of late marriage. Irish fertility and reproduction, abetted by early marriages and the subdivision of farms, deepened the naturalness of the countryside and, for some, further impeded improvement.[42] Both Leadbeater and Doyle warn against early marriage in a broadly Malthusian mode, working to instil 'preventive checks' against the threat of a continually rising and unsustainable population, and thereby the

40 See C. J. Woods, 'Blacker, William'. *Dictionary of Irish Biography*, ed. J. McGuire and J. Quinn (Cambridge, 2009).

41 C. Ó Gráda, *Ireland: A New Economic History, 1780–1939* (Oxford, 1994), pp. 5–13.

42 Doyle proposed an emigration scheme to relieve population. See M. Doyle, *Hints on Emigration to Upper Canada: Especially Addressed to the Middle and Lower Classes in Great Britain and Ireland* (Dublin, 1834).

kicking in of the more ruthless 'positive checks' of famine and war.[43] Sexual restraint, underpinned by a bodily discipline of plain clothes, simple diet, and temperance, was deemed essential to the control of population in Ireland, which itself would result from the instilling of clear distinctions between nature and society, and thus the creation of an evidently humanized social order.

Irish demographic expansion was certainly a factor in propelling the effort to communicate improvement as exhaustively as possible. No one was to be omitted from improvement: women, children, and even animals were all—in one manner or another—addressed by improvement in these years. The diminished scale of post-Union improvement is in part attributable to the fact that many improvement writers were themselves engaged 'on the ground' in the Irish countryside: Leadbeater was an active reformer in her local Kildare community and, it seems, worked as a kind of unofficial agent on the estate of Melissina Trench; Hickey had long immersed himself in agricultural debates and societies in Wexford, where he also established an agricultural school; and, as noted above, Blacker was employed as an agent for several large Ulster landowners. These hands-on experiences made for a certain pragmatism, in which real conditions could at times trump ideological considerations. There was, for example, no denying the necessity of incorporating both women and the very poorest of rural Catholics within improvement, or of the urgency of fully encompassing them all in various progressive schemes. The explicit acknowledgement of the role of women and poor Catholics to the possibility of progress is simultaneously a recognition that social structures, and the kinds of distinctions that underpinned them, were in a process of transformation. It is also an acknowledgement that change might well pursue a different course—one motivated by poor and middling Catholic men and women—to that envisaged and hoped for by improvement writers.

Leadbeater's *The Pedlars* addresses the relationship between gender and improvement in the most direct of terms. In commenting on the improving activities of a particular woman, the question is posed, 'but what can one person, and that one a woman, do?' in the larger context of female powerlessness. In outlining the already thoroughly gendered nature of improvement in Ireland, the reply is direct and unambiguous: 'Yet she, and such as she, are daily improving' the condition of the rural poor 'by employing them, by encouraging them to send their children to schools, to put part of their earnings into

43 An exception is William Richardson, who strongly argued against Malthus's claims regarding the negative effects of Irish demographic expansion. For an account of Richardson's response to Malthus, see Blackstock, *Science, Politics and Society*, pp. 114–30. There are other distinctively non-Malthusian strains in Irish improvement, particularly the encouraging, rather than the excoriating, of a potato diet.

savings banks, and to keep their horses and themselves clean'. Such seemingly small and possibly trivial acts—even, and perhaps particularly, when directed by women—will cumulatively contribute to the complete and 'great alteration' of Ireland as whole: 'By going on in this way, though it may not show much at first, there will be an improvement gradually, for I can myself perceive a great alteration for the better since I was young'.[44] An improved society was not of course to represent any kind of enlightened feminism, liberated from the forces of nature; if anything, it would serve as an effective restraint on any nascent feminism. But it was an explicit acknowledgement of the necessary role of women in the creation of modern commercial culture both in the material sense of actual labour and the more symbolic one of civilizing and humanizing a seemingly patriarchal nature and, by extension, the state as a whole.

Much nineteenth-century improvement began with the assumption that progress could only be achieved by means of an enlightened landlord instigating progressive projects, and thereby infusing both Whig and modestly commercial values, across his estates. Indeed, that assumption endured in various forms throughout the history of Irish improvement. But it did run up against circumstances—demographic, religious, political, and economic—that forced a reluctant democratizing of improvement in the post-Union years. That necessary democratizing of improvement, embracing women and poor Catholics as agents in their own right, resulted from an implicit acknowledgement of the continued development of a middle-class Catholic culture in the post-Union years. In fact, early nineteenth-century improvement is notably shaped, and defined, by the recognition of a burgeoning Catholic culture. While definitively Whiggish about numerous modernizing projects, post-Union improvement was also notably ambivalent about commercial and urban development, which it associated with Catholic interests. Indeed, a legacy of improved efforts to socialize an Irish Catholic nature was to provide a structure and vocabulary in which Irish nationalism would come to articulate some of its objectives. If nineteenth-century improvement had been in part shaped by anxiety surrounding an emerging vocal and politically organized Catholicism, one of its consequences would be the shaping of nationalist conceptions of the state.

44 Leadbeater, *The Pedlars*, pp. 125–6.

# 2

# Palmerston's Conquest of Sligo

*David Brown*

The third Viscount Palmerston (1784–1865) left his mark on the Victorian period as a prominent British statesman (notably his tenure of the Foreign Office, 1830–4, 1835–41 and 1846–51, the Home Office 1852–5, and his two terms as Prime Minister, 1855–8 and 1859–65). His frequently overlooked role as a landlord in Ireland, however, shows in important ways how his Whig instincts for 'progress' (broadly defined) manifested themselves in terms of social and economic, and what can now be considered environmental, change. As this essay will show, Palmerston's career as an Irish landlord represents an interesting case study of environmental 'conquest',[1] in which nature—including also human nature—was seen as something to be mastered, controlled, and even improved.[2] As an initially reluctant, or hesitant, landlord, who did not visit his estates in Dublin and, more substantially, those in Sligo during the first six years of his proprietorship, Palmerston ultimately set out to transform their physical environment in order to advance his broader socio-political and economic goals of making his Irish estates profitable and secure. Alongside his battles with religious groups, government agencies, and local notables in pursuit of the reform of his estates, there is also an important story to be told about how his conquest of nature suggests the ways in which the Irish landscape was, and could be, understood or regarded by this particular Victorian politician and how this reflected prevailing attitudes.

1 Here I borrow the idea of a 'conquest of nature' from David Blackbourn's book of that title in which he offers an analysis of German attempts to control the physical landscape—and especially waterways—over the course of the last two-and-a-half centuries. See D. Blackbourn, *The Conquest of Nature: Water, Landscape and the Making of Modern Germany* (London, 2006).

2 We may note Blackbourn's observation that 'Germans distinguish between "nature" (a cultural projection of human ideas and emotions) and "nature in itself" (the complex projection of life-forms on earth, which includes humans). When I write about the making of the modern German landscape, it is in this double sense. The two meanings complement each other' (ibid., p. 14).

## I

Palmerston rarely referred directly in his correspondence and notebooks to ways in which his view of nature underpinned his approach to estate management. A rare example—if it is indeed an example—is a letter from 1843 in which he told his brother:

> I have been busy reading books on agriculture & horticulture & trying to acquire some knowledge on those matters which are now become sciences. If one does not know something of them oneself one can never hope to get one's estate or garden well managed. I have let all my farms at Broadlands[3] that were out of lease, & tolerably well in spite of the badness of the times. I had a shocking set of bad tenants, but have got rid of most of them, & have brought in people with skill & capital. Our new gardener does pretty well & understands the theory of his department, but he is a Methodist & goes preaching about the country every Sunday, & I fear he thinks too much of his sermons to be very successful in his garden. I must try to put a stop to his preaching.[4]

This was particularly *au courant*, Palmerston's interest in agricultural 'sciences' being piqued in the early 1840s just as work in the field was expanding and acquiring a more systematic and professional character.[5] Palmerston's acknowledgement of the need to understand horticultural and agricultural sciences, even on this relatively mundane and domestic level, points to ways in which they were connected with ambitions of a more economic, political, or moral character. The occasional notebooks which Palmerston kept in the late 1820s, and which contain some short remarks on various questions, but particularly on matters relating to Ireland, reveal something of his thinking in this direction. As well as noting the social tensions he had observed there, he also scribbled one short entry which might be called an environmental comment on the country: 'England might be a great & powerful & prosperous country without Ireland; Ireland never could be a considerable state without England: This is not a Question of Patriotism or national spirit; it is a Question of Geography & Statistics'.[6] Those questions of geography

3 Palmerston's country seat in Hampshire.

4 Palmerston to William Temple, 29 May 1843, Palmerston Papers, Hartley Library, University of Southampton (hereafter PP), GC/TE/306. I am grateful to the University of Southampton Library for permission to quote from the Palmerston Papers.

5 See, for example, S. Wilmot, *'The Business of Improvement': Agriculture and Scientific Culture in Britain, c.1770–c.1870* (Historical Geography Research Series, No. 24, Bristol, 1990), p. 16.

6 Palmerston's notebook containing newspaper cuttings, anecdotes, notes on Irish affairs,

and statistics were ones to which his contemporaries frequently found mainly discouraging answers. Archdeacon O'Rorke, for example, writing towards the end of the nineteenth century, suggested in his history of Sligo:

> When the traveller, on the way from Sligo, reaches Cooldruman rising ground, and looks northward, he is chilled by the change of scene, finding before him a bleak, bare, cheerless country, instead of the rich smiling landscape through which he has been passing; so that it was not altogether the spleen that occasioned Carlyle's exclamation at Cooldruman, 'Lord Palmerston's country—a dingy, desolate looking country. Would we were well out of it!'

The true picture is perhaps more mixed: Carlyle, touring the country in 1849, had actually singled out Sligo as a highlight—the country, he noted, 'alters here', with a 'beautiful view of the sea, Sligo bay, with notable mountains beyond', judging it 'decidedly a pleasant region, with marks of cultivation everywhere, tho' still too treeless'.[7] (The question of trees, or the lack of, is important.) Palmerston's estates in Sligo—covering an area approximately six miles long and two miles wide between the coast and the Benbulben range of hills—was the most significant of his Irish land holdings. He owned land around Dublin too, but it was to his apparently cheerless Sligo estate that he devoted most attention. But whether Sligo and Palmerston's estates were beautiful or dingy, as O'Rorke also observed:

> When it came into the hands of Lord Palmerston it was very much in the state in which Nature formed it: without houses worthy of the name, without cultivated land to speak of—a mere patch here and there, for potatoes, barley or oats—and with two thousand Irish acres of bog, abandoned, in the less sunken spots, to ground game, and, in the swamps, to the snipe, the curlew, and the long-legged crane.[8]

Certainly this Sligo estate conformed to the broader general profile of the region, dominated by small farming in which 'close-knit communities eked out a living on a combination of tillage, fishing, spinning and weaving'.[9] Census returns for 1821–51 show a fluctuating population for the province of Connacht of between *c.*1 million and *c.*1.5 million people, although by 1851,

1827–*c.*1832 (bound volume), PP, BR 22(ii)/13 [p. 77]. Undated, but most likely 1829, given the dates of surrounding entries.

7 Quoted in T. O'Rorke, *The History of Sligo: Town and County* (2 vols., Dublin, [?1890]), i. p. 38.

8 Ibid., ii. pp. 29–30, 32.

9 C. Ó Gráda, *Ireland: A New Economic History, 1780–1939* (Oxford, 1994), p. 34.

in the aftermath of the Famine, the figure of just over 1 million was the lowest of that period, ending, as might be expected, a period of gradual population growth. There was, over the first half of the century, a modest decrease in the level of agricultural employment and a corresponding increase in work in manufacturing, trade, and 'other' occupations—but overall the region retained its overwhelmingly agrarian economic and social character (roughly between two-thirds and three-quarters of employment was in this sector over this period) and there is even evidence of deindustrialization over the slightly longer term.[10]

Palmerston's first visit to the island was not made until 1808, six years after the death of his father when he had inherited the family estates there. Up until this point Palmerston's understanding of Ireland drew on little more than the popular prejudices he had encountered in Edinburgh, as a student at the university, and in novels such as Maria Edgeworth's *Castle Rackrent* (1800), an apposite and cautionary tale which his father had encouraged him to read as part of his studies.[11] By the later 1820s, Palmerston had acquired a better understanding of Ireland from personal experience. On visiting the country Palmerston found it unsurprising that the 'imperious' English approach to the government of Ireland had generated ill-will and tension, rendering a harmonious state of affairs there unattainable. He likened the situation to the ruthless suppression of Anglo-Saxon England by Norman invaders generations earlier.[12] His response was to engage in a series of progressive measures to assist the physical and moral improvement of tenants living on his Irish estates, and by extension the land on which they lived and worked.

Driven by the economic need to make his estates viable and profitable in order to off-set debts and expenses inherited and incurred in England, Palmerston also sought to alleviate poverty on his estates by means of reforms and improvements that were designed to make those estates self-sufficient and productive. As Palmerston wrote to his sister on the occasion of his first visit to Ireland in 1808, 'there is a great deal, I may almost say, everything, to be done', and though the land was 'wholly unimproved' he was confident that 'almost all the waste ground or bog is capable of being brought into cultivation, and all the arable may be rendered worth three times its present value'. Palmerston listed his priorities for his estates as first to address deficiencies: to repair the parish church and build or establish schools and roads; to get rid

10 See T. E. Jordan, ed., *The Census of Ireland 1821–1911: General Reports and Extracts*, I (Lampeter, 1998) for data from 1821, 1831, 1841, and 1851. See also C. Ó Gráda, *Ireland before and after the Famine* (2nd edn., Manchester, 1993), pp. 42–4.

11 D. Brown, *Palmerston: A Biography* (New Haven, CT, 2010), p. 27.

12 PP, BR 22(ii)/13, pp. 73–6. Palmerston's comments are deliberately satirical and his protest that in these references 'I am *of course* not now speaking of Ireland' were clearly ironic.

of middlemen wherever possible. Second, he proposed to set about improving his estates and their management: 'I mean to introduce a Scotch farmer, to teach the people how to improve their land; to establish a little manufacturing village in a central part of the estate, where there are great advantages of water and stone; and to build a pier and make a little port near a village ... called Mullaghmore'. Schools and roads Palmerston identified as the 'most important points at present', addressing fervent demands from his new tenants for such building works, as well as 'no petty landlords' or middlemen.[13] Although Palmerston might have been a benevolent landlord (and that is not to suggest that he was a model landlord), subscribing regularly to relief committees and agreeing to adjustments to rentals where necessary, he was always mindful of the pressing need to make his estates pay. One senses in his distribution of charity in the early years of his landlordship in Sligo that he saw the need to increase the productivity of the land as both an economic imperative for himself as well as a means of improving the moral welfare and well-being of his tenants. This meant regaining control of both the people and also the land on which they lived.

There were man-made as well as natural factors limiting the productivity of the land. The estate was, where habitable, densely populated, and obstacles to improvement existed at every turn in the form of land leased to middlemen under the rundale[14] land system.[15] Palmerston used the occasion of William IV's death in 1837 to accelerate reforms to the system of land-holding on his estates. Many of the middlemen holding leases on Palmerston's land held those leases for the life of the king and Palmerston was urged by his agent Joseph Kincaid to use this as the grounds on which to press ahead with plans for 'squaring' the land, that is to amalgamate many small (subsistence) holdings into larger, more efficient, farms.[16] Although initially wary of the human cost of such reforms and reluctant to impose the system at first, Palmerston increasingly saw it as a necessary prelude to broader changes to the landscape, and his estate was largely 'squared' by late 1846.[17]

13 Palmerston to Elizabeth Temple, 12 Sept. 1808, H. L. Bulwer, *The Life of Henry John Temple, Viscount Palmerston* (3 vols., London, 1870–4), i. pp. 85–8. See also K. Bourne, ed., *The Letters of the Third Viscount Palmerston to Laurence and Elizabeth Sulivan, 1804–1863* (RHS/Camden Society, 4th ser., xxiii, London, 1979), pp. 104–5.

14 Under the rundale system, individual plots of disconnected land were rented as single leases meaning tenants were unable to consolidate holdings and achieve potential economies of scale.

15 O'Rorke, *History of Sligo*, ii. p. 32; D. Norton, 'On Lord Palmerston's Irish Estates in the 1840s', *English Historical Review*, cxix (2004), p. 1256.

16 On the squaring of land in this period, see D. Norton, 'On Landlord-Assisted Emigration from Some Irish Estates in the 1840s', *Agricultural History Review*, liii (2005), pp. 27–8 and Norton, 'On Lord Palmerston's Irish Estates', pp. 1256–7, 1268–9.

17 Norton, 'On Landlord-Assisted Emigration from Some Irish Estates in the 1840s', p. 28.

## II

One can glimpse something of the scale of the physical reforms planned and undertaken in a snapshot from 1826. Alexander Nimmo, an engineer much employed by Palmerston, reported favourably on works that summer, having got 'a good deal done towards your various improvements'. Palmerston's own summary at the end of the report listed:

> 1. Bent planted on sand hill doing well.
> 2. Embankment at Bunduff Marsh has answered & Marsh is grown firm & pasturage much improved.
> 3. Road to Castlegall from sluice opened but not quite finished.
> 4. Road from Grange to Lyle in great forwardness.
> 5. Embankment for turning Cartron River across sands finished but must be strengthened before flat sand behind it is cultivated.
> 6. Operations at Grellagh going on well.
> 7. Road to Limestone at Derrylighan finished. Lime burnt for 6d a barrel, will not cost more than 5d in future.
> 8. Road across bogs to Cliffony marked out & drained.
> 9. Pier at Mullaghmore can be prolonged as proposed by me—will send an enlarged plan.
> 10. Cost of what has been done on pier & of what remains unfinished—147£ due to England for what has been done.
> 11. Has had conversation with the distiller who will meet me at Sligo a good flour mill at Grellagh advisable.[18]

Such things did not come without substantial financial outlays on Palmerston's part. A list of expenditure for the years 1825–7 detailed: a new harbour at Mullaghmore (costing approximately £3,133); housing (almost £500); building of embankments and planting of bent (over £973); cultivation of bogland (more than £975); spending on housing, schools, and an inn at Cliffony (about £1,700); road building (£300); developing a lime works (almost £200); and further sums on other investments in land and building which brought the total cost to £7,354.[19]

Nimmo was a key influence. Born and educated in Fife, he had a formidable reputation as a surveyor and engineer both in Scotland (where he had advised commissioners for roads and bridges in the Highlands in 1806) and in Ireland (where, on Thomas Telford's recommendation, he had been employed between 1810 and 1813 by a government commission to survey bogs with a

18 Alexander Nimmo to Palmerston, 14 July 1826, PP, BR 147/10/41.
19 'Abstract of Expenses of Various Works in Ahamlish in 1825–26–27', PP, BR 139/24.

view to assessing their potential for drainage and cultivation). Nimmo's maps and recommendations for land reclamation in Ireland, as well as proposals for new harbours and other coastal improvements, were published in 1814 as part of the commissioners' fourth report, and then in 1814–15 Nimmo embarked on a tour of public works in France, Germany, and the Netherlands which informed his subsequent work in Ireland on harbours and coastal environments for the fisheries board. It has been estimated that more than thirty harbours or piers were eventually built under Nimmo's direction.[20] Among these was the harbour at Mullaghmore.

The harbour at Mullaghmore stands as one of the most significant and tangible examples of Palmerston's approach to the reform of his Sligo estates. Nimmo appears to have come to Palmerston's attention by at least October 1820, when he was advised that Nimmo had just completed a survey of the west coast of Ireland and had identified Mullaghmore as 'an improvable and useful harbour', recommending that Palmerston approach the Commissioners for financial aid to help undertake the work.[21] As well as being a suitable site from a practical point of view, offering shelter from storms, a good depth of water, and proximity to local building materials, Mullaghmore was close to prime fishing waters with healthy stocks of turbot and herring in particular, and white fish generally.[22]

Palmerston endorsed the plan enthusiastically, and although financial mismanagement meant the Commissioners could not finance the plan Palmerston undertook to pay for the works himself and seek reimbursement in the future, even though this reimbursement seemed hopeless.[23] He anticipated that the harbour project would underpin a wider economic development based on exploiting the region's natural resources, which might involve laying down a railroad to the end of Loch Erne, making Mullaghmore 'the exporting and importing harbour for a large tract of very fertile country lying on the banks of that lake'.[24] Adverse weather conditions and unreliable supervisors of works slowed progress,[25] but still the first stage of building was near completion by late 1826. At this point Palmerston wrote to the Fisheries Board, advising them that he would assume sole responsibility for financing the scheme, while also articulating the importance of developing the fishing industry on the west coast of Ireland: the 'coast is thickly peopled, much too

20 Roland Paxton, 'Nimmo, Alexander (1783–1832)', *Oxford Dictionary of National Biography* ([2004] Jan. 2008 online edition).

21 G. Swan to Palmerston, 30 Oct. 1820, PP, BR 147/6/32.

22 J. McGowan, *In the Shadow of Benbulben* (Manorhamilton, 1993), p. 198.

23 Memorandum by G. Swan, Sept. 1822. Palmerston approved the plan on 24 Nov. 1822. PP, BR 147/7/42.

24 McGowan, *Shadow of Benbulben*, p. 199.

25 For accounts of progress on the harbour, see PP, BR 147, *passim*.

thickly for the purposes of agriculture, and in the absence of manufactures the abundant fishing grounds which are to be found along the shores afford the readiest means of calling into activity the industry of the sea coast population'. Recognizing, however, that the best fishing grounds were 'some distance from the shore' and beyond the reach and capacity of the row boats currently in use Palmerston proposed that the Board provide two twenty-ton vessels, the people of Mullaghmore being themselves too poor to provide such boats for themselves.[26] It is unclear whether the Fisheries Board did supply the boats, although a cutter, the *Benbulben*, was in operation by the summer of 1829.[27] The harbour was ready by spring 1828 and Palmerston himself drafted an advertisement for the Sligo and Fermanagh newspapers highlighting the trading potential of the new facility.[28] Early accounts of the expansion of the fisheries off Mullaghmore were encouraging, as were reports that the harbour was regularly receiving significant cargoes.[29] In addition to stimulating local fishing and supplementing local agriculture, reports from Palmerston's agents suggested that the harbour also appeared well-placed to provide fresh stimulus to the local economy, the timber trade in particular having benefited.[30] The kelp industry also grew thanks to the harbour, by providing the means to sell it well beyond its erstwhile reliance on local markets.[31]

Despite these auspicious beginnings, the harbour's capacity was soon found wanting: it was still not sufficiently accessible to larger vessels, for which it would be necessary to extend the pier.[32] Again Palmerston invested heavily (a further £1,000 in the late 1830s) and with some assistance from the Board of Works the harbour was extended.[33] The result, by the mid-1840s, was judged a great improvement.[34] Harnessing the potential of the coastline had thus underwritten a significant restructuring of the local economy, which both physically built on that landscape and drew from—in terms of things like fish and seaweed—its natural resources. This might be read as an important

26 Palmerston to Henry Townshend [secretary to Fisheries Board], 28 Oct. 1826 (draft copy), PP, BR 147/10/63.

27 Henry Townshend to Palmerston, 7 Nov. 1826, PP, BR 147/10/64; James Walker to Palmerston, 4 July 1829, PP, BR 145/2/30; E. Nicholson to Palmerston, 16 and 25 July 1829, PP, BR 145/2/28, BR 150/9/29; H. Townshend to Palmerston, 2 Nov. 1829, PP, BR 145/2/5.

28 Draft advertisement by Palmerston, Mar. 1828, PP, BR 145/1/10.

29 For example, Herbert Clifford [Insp. Comr. Sligo Coastguard] to Palmerston, 3 Dec. 1831, PP, BR 145/4/14.

30 See, for example, John Lynch to Palmerston, 5 Dec. 1831, PP, BR 145/4/15.

31 McGowan, *Shadow of Benbulben*, pp. 217–18.

32 James Walker to Palmerston, 24 Aug. 1835, PP, BR 145/7/34.

33 J. Kincaid to Palmerston, 21 June 1837, PP, BR 145/9/14/1.

34 See, for example, John Lynch to Palmerston, 25 Oct. 1845, PP, BR 146/7/48.

example of Palmerston using the environment to achieve his economic and political goals.

If the harbour was a success, alone it was not enough to alleviate poverty on Palmerston's estates. He insisted that public works schemes were essential forms of relief as well as a means of improving the infrastructure of the local economy. Any improvement to roads, canals, harbours, public buildings, bridges, and land would invariably alleviate hardships by underpinning economic growth and increased employment. Such investment, argued Palmerston, promised to 'confer great & lasting benefits upon Ireland', and, as he had outlined in an (unsent) letter to the Committee of Relief for the Town of Sligo and the Union of St John in 1822, it was incumbent on those with the means to do so to 'accelerate' society's 'march in internal improvement'.[35] Palmerston had recognized early on the need for improved roads and rail links to and across his Sligo estate, and his tenants, too, were clearly anxious for better transport connections. Partly at his own expense and partly with public subsidy, the towns and farms on Palmerston's estates were gradually linked with each other, with the coast, and with markets in the rest of Ireland.[36] In a very crude sense, building houses and laying roads is a clear example of changing the landscape, altering nature, conquering the terrain, and should form part of an environmental assessment of Palmerston's record. Yet, while infrastructure was developed, there remained the problem of the land itself.

As O'Rorke had observed, Palmerston's estates were dominated by bog. Across the country, the necessity of reclaiming bog had long been recognized and it had been a government priority since the early eighteenth century, the need becoming ever more urgent as the population expanded in the nineteenth.[37] Palmerston's own calculations, based on figures produced sometime between 1818 and 1821, estimated the area of uncultivated bog on his estates at almost 2,200 acres.[38] Palmerston recognized that the tenants themselves were not capable of undertaking this transformational work and he set about draining and reclaiming these lands himself. He turned again to Nimmo to prepare schemes that would allow him to extend the cultivable land on his estates that it might then be let as a series of ten-acre holdings to families living on the coast. Not content simply to drain and populate the land, Palmerston was also alert to new methods of stabilizing agricultural land, and he urged that screens of ash, larch, fir, and oak be planted where

35 Palmerston to Wm. Urwick, 22 June 1822 (not sent), PP, BR 147/7/16/2.

36 See, for example, reports on works submitted to Palmerston by agents, esp. PP, BR 147.

37 P. Foss and C. O'Connell, 'Bogland: Study and Utilization', in J. W. Foster and H. C. G. Chesney, eds., *Nature in Ireland: A Scientific and Cultural History* (Dublin, 1997), p. 192.

38 List of 'Quantity of Uncultivated Bog on my Sligo Estate', n.d. [1818–21]. The total figure given is: '2184A, 2R, 29P', PP, BR 147/6/48.

necessary.[39] As already noted, the destructive effects of wind and sand on the landscape had a clear human as well as agricultural impact.

The uneven, perhaps unpromising, state of the land on Palmerston's estates is in many ways a product of the rapidly (in a geological sense) changing nature of this landscape on Ireland's west coast. Don Cotton's work on peat deposits indicates that at times in the last thousand years terrain now undersea was once land.[40] In the nineteenth century, this had created a land of poor quality soil, and large swathes of bog sitting, in the main, on a subsoil of freestone gravel.[41] Winds from the Atlantic did further damage to the agricultural potential of the area, blowing sand along the coast and turning what green land there was into something resembling a beach. By Palmerston's own estimate, something like 600 acres had been destroyed by the wind and sand and, as O'Rorke put it, 'the people felt themselves powerless, and resigned themselves to the loss, as to the inevitable'.[42] Mastering the land, therefore, was not only about improving crop yields and increasing revenues, but also about overcoming this fatalistic attitude towards the natural environment. One contemporary account of the county evoked the extent of this problem:

> we almost waded through this wilderness, saw no house, and were wondering how much farther we had to go when suddenly the constable stopped opposite to what appeared to be a thatched potato pit, nearly covered with sand. We went round, saw an opening in the thatch, and the end of a ladder protruding; by this means we descended and dimly descried the poor man and his family, by means of the light given by the hole in the roof and by the chimney. His recovery was slow, and when able to be moved, he was at once carried off to his neighbours in the adjoining townland, as by that time the sand had completely covered the roof of his former dwelling.[43]

Many letters between Palmerston and his agents discuss the sowing of bent, a type of grass that would help avoid loose and fragile land being blown away.[44]

39 Memorandum by Palmerston for Mr Nimmo: 'Instructions for surveying the bogs on my estate in Sligo with a view to drain & reclaim them', 12 Oct. 1824 (copy), PP, BR 149/6/1/4.

40 D. Cotton, 'The Environmental Setting', in McGowan, *Shadow of Benbulben*, pp. 259–67 (here p. 264).

41 O'Rorke, *History of Sligo*, ii. p. 30.

42 Ibid., ii. p. 32.

43 McGowan, *Shadow of Benbulben*, p. 161.

44 See correspondence between Palmerston and his agents collected in PP, BR 145, BR 146, BR 147, BR 148, BR 149, BR 150. The earliest listed letter dealing with bent planting is from James Walker, 26 May 1825 (BR 147/9/28).

In 1826, he planted 140 acres of bent, which brought the total area he had thus sown close to 600 acres.[45] Yet Palmerston also exploited his Foreign Office contacts to investigate other methods. In 1835, he had the British government's consul in Bordeaux send over, at Palmerston's own expense, twelve bags of the seed of the *Pinus maritima*, which had apparently served well on the sand hills of Bordeaux.[46] By doing so, Palmerston positioned himself as a progressive and innovative landlord; and, as one commentator noted in 1854, 'this has been by far the most extensive ... experiment of the kind yet undertaken'.[47] Within a month of receiving the Bordeaux tree seed, Palmerston was approached by Robert Stevenson of Edinburgh who wanted to know more of Palmerston's attempts to use 'sea fir to stop blowing sand',[48] while in the spring of 1836 Palmerston wrote to Daniel O'Connell, who had also enquired after the experiment, and urged him to try the Bordeaux pine as well, keen to see whether the seed might receive a 'fairer trial' further south.[49]

Palmerston also used his estates as a nursery: in 1832, an advertisement for a sale of 'Forest trees to be sold on Lord Palmerston's Nursery' listed 182,000 trees for sale—including Ash, Elm, Canada Poplar, Huntington and Redford Willows, Birch, Laurel, Scotch Fir, Larch, Alder, Oak, Norway Spruce, and White American Spruce[50]—and a similar advert the following year for about 300,000 trees.[51] The cultivation of trees is particularly significant in terms of transforming, or re-forming, the landscape. Early modern deforestation of Ireland, in part for economic reasons but also as a means of destroying the shelter of those resisting colonial conquest, had left the country largely 'treeless' by the nineteenth century. Attempts since the late eighteenth century to encourage reforestation notwithstanding, even by the late 1880s a French visitor to the country would observe that the 'most striking thing on first sight of the Irish landscape is the total absence of trees of any kind', those that were to be found being 'only in private parks'.[52]

The techniques employed by Palmerston gradually bore fruit, or rather vegetables and grain. By 1826, for example, thirty acres of previously unworkable

45 McGowan, *Shadow of Benbulben*, p. 165.

46 T. G.[?] Scott to Palmerston, 17 Nov. 1835, PP, BR 145/7/46.

47 J. Fraser, *Hand Book for Travellers in Ireland* (Dublin, 1854), quoted in Desmond Norton, *Landlords, Tenants, Famine: The Business of an Irish Land Agency in the 1840s* (Dublin, 2006), p. 27.

48 Robert Stevenson to Palmerston, 16 Dec. 1835, PP, BR 145/7/49 (Palmerston's endorsement quoted).

49 Palmerston to Daniel O'Connell, 30 Mar. 1836, PP, BR 145/8/15.

50 McGowan, *Shadow of Benbulben*, p. 165.

51 Norton, *Landlords, Tenants, Famine*, p. 27.

52 Paschal Grousset quoted (from 1887) in E. Neeson, 'Woodland in History and Culture', in Foster and Chesney, *Nature in Ireland*, p. 142. On deforestation and reforestation in earlier periods, see ibid., pp. 140–1, 145–6.

bog were now producing potatoes, turnips, and rape.[53] In 1835, William Bowles, one of Palmerston's agents, reported on the universally improved circumstances prevailing on his estates. Mullaghmore, he observed, was 'wonderfully altered for the better since I saw it last', thanks to the new pier, which was almost finished, a 'very good corn store full of fine oats', 'a small neat inn, and several cottages of better description, besides the coast guard buildings which are very comfortable'. Elsewhere the picture was equally pleasing: the bent planting had succeeded and Bowles 'was agreeably surprized to see covered with *bent* throughout, and in many places where it was earliest planted (five and six years ago) a fine bottom of natural grass and clover beginning to clothe the whole surface, and there seems not the least reason to doubt that in a very short time you will have a fine sheep walk there'; he surveyed too 'the reclaimed bog' which now had 'a very fine crop of oats', praised the rail road, which 'saves an infinity of labour', and even suggested reclaimed land 'now let, seem[ed] rather overstocked with cattle, and getting *out of heart*'. A 'thriving Nursery Garden' supplied the neighbourhood with trees, including a local landlord who was following Palmerston's example, while 'the great number of neat villas and comfortable houses which are gradually rising up on all sides of the town' testified to the success of the improvements. Bowles dismissed reports of hardships as having been 'exaggerated by the priests and others whose purposes it suited to magnify the distress, that they might have the whole credit of relieving it'.[54]

A report on the estate by Palmerston himself in 1858, even after the effects of the Famine the previous decade, still painted a very positive picture, but also pointed to the extent to which this was an ongoing battle: more trees, more bent, more reclamation of land was required to maintain the rate of improvement.[55]

## III

It is important also to notice the ways in which Palmerston's transformation of the landscape left an important legacy, developing new markets, industries, and opportunities. In the later nineteenth century, the works around Mullaghmore, for example, had opened up the possibility for the development of the mineral waters of the area to be promoted as health-giving in ways 'fully equal' to those of Vichy, Ems, Carlsbad, Wiesbaden, and any other continental European spa town.[56] The town had already begun to

53 McGowan, *Shadow of Benbulben*, p. 165.
54 William Bowles to Palmerston, 22 July 1835, PP, BR 145/7/32.
55 Palmerston to Emily Palmerston, 14 Aug. 1858, PP, BR23AA/1/35.
56 Mineral Analysis, 1874, by the Duke of Connaught, in McGowan, *Shadow of Benbulben*, pp. 268–9.

be marketed as a bathing place,[57] while in another attempt to draw out the benefits of the region's mineral resources, Palmerston had built lime kilns in the late 1820s that were soon producing lime at a decent profit.[58]

Palmerstonian reforms of the physical landscape had taken three broad forms: first, innovation, such as the building of a harbour at Mullaghmore and laying the foundations of a more secure fishing industry; second, improvements to existing towns and associated facilities; and, third, the reclamation of erstwhile unusable land. It is in this sense that Palmerston undertook a 'conquest of nature' in Sligo. Palmerston concerned himself, too, with the general moral, spiritual, and educational welfare of his tenants, and while that is an important part of the environmental story it is not something that can be discussed explicitly within the confines of this chapter.[59] Taken together, however, these improvements and reforms represented a holistic attempt to change, and improve, this quarter of Ireland. And so while it would be misleading to represent Palmerston as a model landlord, his programme of improvements during the 1820s, 1830s, and 1840s did effect significant changes to his Sligo estates and transformed the landscape.

By 1848, even in the wake of the disastrous effects of the Famine, Palmerston's estate seemed to stand out as a model for others to follow. A lengthy letter from a J. R. Bertolacci to Sir Charles Trevelyan in that year surveyed the condition of various English landlords' Irish estates and singled out Palmerston's lands for especial praise: 'there is a farm on the property of Lord Palmerston', he wrote, which 'is like a star on the face of the country compared to the land I have before described and shows what *could* be done'.[60] Palmerston's tenants, too, seem to have regarded his exertions as having produced beneficial results. In 1840, on the occasion of a visit from their landlord, Palmerston's Castlegarron tenants presented an address observing that in the previous twelve years Palmerston had re-invested in the region almost all of the rental income he had derived from the estate: 'We most certainly attribute such unexampled liberality to benevolence alone, as no adequate remuneration can probably be derived from such an expenditure'. In return they hoped that, 'by our peaceable, industrious, sober, and temperate habits to give your Lordship the most decided proof that your munificence and liberality have not been thrown away upon us, and in acting thus we feel certain we best meet your Lordship's wishes'.[61]

57 See the advert from the *Sligo Journal*, 1836, ibid., p. 196.
58 Ibid., pp. 183, 186–7.
59 For a wider discussion of Palmerston's estate management, see Brown, *Palmerston*, esp. pp. 89–103.
60 J. R. Bertolacci to Sir Charles Trevelyan, 21 Nov. 1848, PP, BR 146/10/18.
61 Address of Castlegarron Tenants, 22 Oct. 1840, PP, BR 146/2/28.

This last point is significant. For in describing Palmerston's 'conquest' of Sligo, it might be argued that he was using a mastery—or conquest—of the physical landscape as a means of shaping society. The economic benefits of the reform of the environment seem pretty clear, but underpinning that was a desire to reform the people living on that land: Palmerston's activities were driven by a determination to bring a Whiggish view of the exercise of power—for the people, but not of them, and infused with some measure of progress—to bear on the estate. Writing to his brother in late 1827, he reported on 'a three weeks trip to Sligo where I found my improvements going on well', and expressed his hope that as a result of the reforms of the landscape he would 'find my people in a few years somewhat resembling a civilized race'.[62] As Helen O'Connell demonstrates elsewhere in this volume, 'improving' the environment and agricultural productivity was something indivisible from seeking social and moral advances in the condition of the people, binding landlords and tenants in a shared enterprise to modernize the country according to a particular English vision of progress.[63] In pursuing this sort of top-down management Palmerston might have been substituting one sort of dependency for another, but the restructuring of the terrain was integral to Palmerston's wider political ambitions for his estate and its tenants.

This argument risks painting a rosy image of Palmerston's record. With any conquest there are winners and losers. The road building and reorganization of the land led to the destruction of a number of villages, and to emigration, as land holding was rationalized. Whether this was positive or negative depends to some extent on where one stands in relation to that dislocation, but for all the new building and new opportunities this conquest of the land did lead to some losing their homes.[64] Moreover, these land reforms affected biodiversity and sustainability. Here the evidentiary record is less straightforward and clear. It is difficult to know how far Palmerston's efforts to change the land impacted on the Famine of the 1840s directly, but the greater stability of the land had not necessarily been a prelude to greater crop diversity, or a noticeable difference in the county's ability to withstand the devastating effects of the potato blight. While the harbour in Mullaghmore and the reclamation of bog had created more cultivable land and a more diverse food chain, ultimately, as local newspapers were soon reporting, whatever attempts Palmerston had made to diversify the economy, the land was still in many places of low quality and 'poor fishermen were suffering greatly, fish

62 Palmerston to William Temple, 27 Nov. 1827, PP, GC/TE/190.

63 See Helen O'Connell, 'The Nature of Improvement in Ireland', Chapter 1 in this volume.

64 For a vigorous account of these processes, see B. Mac Suibhne, *The End of Outrage: Post-Famine Adjustment in Rural Ireland* (Oxford, 2017).

being scarce, or none caught as formerly'.[65] A report from one of Palmerston's agents in 1846, for example, underlines the problems of an over-reliance on a narrow range of crops and offers an important qualification of positive assessments of Palmerston's reforms of the landscape and agriculture. People were struggling to survive on the 'small quantities of grain they sewed', which in any case were almost exhausted, Palmerston was told, and with no supplementary nourishment from the sea many of his tenants were 'in a state of starvation having nothing to eat but *boiled* Cabbage and a little salt', and, moreover, 'this vegetable has now nearly disappeared'. Unemployment was high and increasingly tenants were dependent on subsidized food supplied by landlords.[66] To quantify the problem: according to one contemporary account, in Mullaghmore, 728 people 'were solely dependent on six acres of corn, having lost all their potatoes, the food on which they mainly existed'.[67] While the land reclamation, stabilization, and improvement had increased yields in the 1820s and 1830s therefore—and indeed would again in the 1850s and 1860s—to a large extent Palmerston had focused on how much, more than on what, was produced and so his estates' viability remained precarious, dependent on favourable environmental conditions.

There is a wide literature dealing with the sustainability and robustness of the Irish economy at this time, and while the picture is a nuanced one reflecting variations by region and class over time, the consolidation of a 'potato economy' by the middle of the nineteenth century meant that Ireland suffered from the blight of that period more severely than other European countries. While pre-Famine agriculture in Ireland had seen increasing productivity, sufficient to sustain a growing population and also underwrite an emergent export trade, as well as some preliminary signs of 'modernization' (for example, in the shift away from rundale holdings), there was, nonetheless, an overwhelming reliance on one staple crop, the potato.[68] It was this that made the effects of the blight so much more severe in Ireland than elsewhere: in Scotland, failure of the potato crop in 1846 and 1847 led to significant numbers of deaths,[69] but the dependence on that food was not, as in Ireland, 'absolute', at least when speaking of the rural poor.[70] Although the potato was not the sole source of

65 See, example, newspaper cutting, PP, BR 148/3/9/2.
66 Revd Malachi Brennan to Palmerston, 24 Oct. 1846, PP, BR 148/3/15.
67 J. Kincaid to Palmerston, 7 Nov. 1846, enclosing cutting from a Sligo newspaper (unspecified), PP, BR 148/3/17.
68 M. Daly, 'Farming and the Famine', in C. Ó Gráda, ed., *Famine 150* (Dublin, 1997), pp. 31–7.
69 C. Ó Gráda, *Black '47 and Beyond: The Great Irish Famine in History, Economy, and Memory* (Princeton, NJ, 1999), p. 16.
70 C. Boylan, 'The Famine', in R. Bourke and I. McBride, eds., *The Princeton History of Modern Ireland* (Princeton, NJ, 2016), p. 405.

nourishment in Ireland, something like one-third of land under cultivation was used for potatoes and it dominated agricultural production in the vast majority of Irish counties,[71] ensuring the blight had catastrophic consequences.[72] In the immediate short term, the potato was a relatively effective form of basic nourishment, but it had significant shortcomings: it was almost impossible to store, particularly in the wet climate of Ireland (making it impossible to build up reserves to set against future failures), it was difficult to transport, and, by the 1840s, production had shifted from the higher-quality varieties ('Black' and 'Apple') to higher-yield but lower-nutritional-value types ('Cups', and, especially, 'Lumper').[73] Austin Bourke wrote of a 'sinister trend towards monoculture' during the later eighteenth and early nineteenth centuries,[74] something which Palmerston's improvements did little to address.

As such it is important to consider the role of 'improvement' in the Famine. In analyses of more recent agricultural development, the Irish Famine of the 1840s has often been singled out as a warning against simplification (narrowing of crop diversity) and associated drives towards agricultural monocultures.[75] While climatic conditions and diseases can explain the timing of crop failures, they are only a partial explanation of the resulting famine; it is just as important to take into account the over-reliance on one crop in a densely populated country.[76] In 1840s Ireland, famine happened because of the potato blight, exacerbated by harsh weather conditions; but it also happened because of problematic government responses (relying on inadequate public works schemes to provide relief during the harshest of winters),[77] prevailing ideas about government responsibility (the dominance of laissez-faire),[78] and religion (including notions that the Famine was a providential judgement).[79]

Bourke was drawn to conclude that the potato was 'no more than the passive instrument of misery and famine' and the real cause of the crisis was

71 Ó Gráda, *Black '47*, p. 19; D. Dickson, 'The Potato and Irish Diet Before the Famine', in Ó Gráda, *Famine 150*, p. 5; A. Bourke, *'The Visitation of God'? The Potato and the Great Irish Famine* (Dublin, 1993), p. 21.
72 Boylan, 'The Famine', p. 408.
73 Bourke, *'The Visitation of God'*, pp. 32–40.
74 Ibid., pp. 20, 67.
75 J. C. Scott, *Seeing Like a State: How Certain Schemes to Improve the Human Condition Have Failed* (New Haven, CT, 1999), p. 268.
76 C. Vita-Finzi, 'Climate and History in the Old and New Worlds', in P. Slack, ed., *Environments and Historical Change: The Linacre Lectures 1998* (Oxford, 1999), p. 34; W. J. Burroughs, *Does the Weather Really Matter? The Social Implications of Climate Change* (Cambridge, 1997), p. 43.
77 Ó Gráda, *Black '47*, p. 36.
78 Ibid., p. 6.
79 On providentialism and its influence on government, see P. Gray, *Famine, Land and Politics: British Government and Irish Society, 1843–50* (Dublin, 1999).

the placing of government and landlord self-interest above that of tenants.[80] This is not to say that Palmerston's estate improvements caused the Famine, but nor did they create a system capable of withstanding crop failure and mitigating its effects. Due to the specific combination of political, religious, and economic as well as environmental reasons, it is clear that Palmerston's conquest of Sligo was in many ways a rather fragile one.

## IV

It is worth turning in conclusion to reflect more broadly on what Palmerston's 'conquest' of Sligo might signify in terms of what drove his attempt to master the physical environment—beyond the obvious economic and political impulses already noted—and whether changes wrought by Palmerston were specifically 'Irish'.

Palmerston's efforts to tame nature were part of a much broader Enlightenment project, and it has been argued recently, for example, that 'the mountains and peat mosses of Scotland became a laboratory for the Enlightenment', which saw understandings of modernization emerge directly conditioned by understandings of the environment.[81] As the Scottish philosopher James Dunbar put it in 1780: 'Let us learn to wage war with the elements, not with our own kind; to recover, if one may say so, our patrimony from Chaos, and not to add to his empire'.[82] Frederick the Great made a similar claim: 'Whoever improves the soil, cultivates land lying waste and drains swamps, is making conquests from barbarism'. Quoting this, David Blackbourn has described Frederick's as 'the authentic voice of eighteenth-century enlightened absolutism'.[83] Thus Palmerston's use of pine trees might have been important in stabilizing shifting sands—and he might have played an important role in introducing this technique to Ireland and Scotland—but he was not necessarily doing anything new in the broader scheme of things: the trees he imported were chosen precisely because they had been doing the job well in France; while Frederick the Great had apparently been trying to use them for this purpose since the 1770s.[84]

While taming nature might have been rooted in wider European traditions, it is suggested that 'improvement', as an ideal, was a central feature of an emerging (early) modern English identity. 'Improvement' in this sense meant

80 Bourke, '*The Visitation of God*', pp. 71–2. See also Boylan, 'The Famine', esp. p. 419.
81 F. A. Jonsson, *Enlightenment's Frontier: The Scottish Highlands and the Origins of Environmentalism* (New Haven, CT, 2013), quotation from p. 2.
82 Quoted in Blackbourn, *Conquest of Nature*, p. 64.
83 Ibid., p. 41.
84 Ibid., p. 42.

positive change, but also gradual and therefore safe (not revolutionary or radical) reform, that could frame national narratives of betterment over the course of the seventeenth and eighteenth centuries and be understood, shared, and celebrated as a distinguishing feature of Englishness. Rising living standards over that period only seemed to confirm that the country was indeed improving: by the late eighteenth century, 'when Britain was becoming the first industrial nation, and distributing its culture with its empire across the world, improvement was no longer a cause that needed defending at home, but a conviction fortified by complacency'.[85] While improvement was inextricably linked with profitability and financial gain, that was frequently understood in terms of increasing the productivity of land and its rentable value, as much, if not more than, a more general meaning of general amelioration and betterment.[86] This chimes clearly with Palmerston's Whiggish notions of seeking improvement to secure economic benefit.

Yet the picture is a little more interesting than this: improvement was also about enlightened government, and social progress; rising living standards only justified and facilitated that. Agricultural and environmental 'improvement' could be seen as a means of anglicizing Ireland, and, by extension, consolidating the union of Britain and Ireland in political terms.[87] While the actual, tangible, benefit of advances in agricultural science in terms of increased yields and productivity might have been open to debate, and while there was some wariness about the desirability of creating a genuinely self-sufficient agricultural population, active engagement with the forces of improvement and innovation were increasingly recognized as a means of consolidating structures of power and government. In the early nineteenth century this had been about creating well-ordered estates, which, as they were improved, underpinned the traditional notion of a compact between landlord and tenant in which increased productivity benefited all parties. By the mid-nineteenth century, the point at which Palmerston was becoming more interested in agricultural science, the landlord-as-expert was increasingly revealed to be landlord-as-amateur, and the literature of the 1840s spoke ever more of the need for the landlord to be simply a facilitator, a provider of capital, within a transformed rural society and not its director.[88] More significantly, perhaps, while 'improvement' was often presented as politically neutral, increasingly this became a motif that could signal a shift away from Catholic hierarchies. As O'Connell argues, the 'displacement of traditional agriculture was supposed

85 P. Slack, *The Invention of Improvement: Information and Material Progress in Seventeenth-Century England* (Oxford, 2015), pp. 1–4 (quotation from p. 4).

86 Of the six broad definitions of 'improvement' in the *OED*, five are concerned with profit, productivity of land, and efficient exploitation of resources.

87 O'Connell, 'The Nature of Improvement in Ireland'.

88 This discussion draws on Wilmot, '*The Business of Improvement*', pp. 24–80.

to stimulate and go hand-in-hand with the displacing of traditional Catholic culture more generally, which was understood to be worryingly naturalized'.[89] While this does not on its own explain Palmerston's protracted struggles with the Catholic clergy on his estates, it does help reframe those contests within a broader framework of a narrative of (English, Whig) 'improvement'.

It is in these contexts of seeking improvement, therefore, that a modern well-ordered estate, which had reclaimed the land from its wild and unproductive condition, could also be linked explicitly to notions of civilization, a demonstrable shift away from an earlier perception of those 'boundless bogs' as 'infamous' and associated with barbarism.[90] While poets and pictorialists eulogized a golden age,[91] landlords such as Palmerston saw the wild landscape as something that could, and should, be tamed. Writers such as Edgeworth, for example, with whom Palmerston was already familiar, in discussing improvement, were, it has been argued, in many ways explicitly rejecting a romanticized view of the natural landscape.[92] The refined and modernized landscapes of improved estates might be seen as a statement about perceived degrees of civilization. It has already been shown how Palmerston linked change with civilization in a social sense, but this ought to be considered against a broader background of how the landscape was read. As Blackbourn has pointed out, eighteenth-century travellers saw in transformed landscapes evidence of progress.[93]

Of course, there is a counter-point here, as Blackbourn himself acknowledges, that created an alternative vision of the ideal 'untouched' (unspoilt) paradise; but to many, Palmerston included, such was a 'dangerous foe'.[94] It might have been that the forestation and reclamation projects were interpreted as symbols of colonization and of the conqueror's wealth and status,[95] but that only confirms the sense that landscape reform was in many respects about negotiating, or

89 O'Connell, 'The Nature of Improvement in Ireland'.

90 See, for example, J. W. Foster, 'Encountering Traditions', in Foster and Chesney, *Nature in Ireland*, p. 26, citing William King (subsequently Archbishop of Dublin), 'Of the Bogs and Loughs of Ireland' (1685).

91 See, for example, the supposed romantic ideal of the 'natural' landscape, as depicted in Lady Morgan's 1806 novel *The Wild Irish Girl*, a narrative inspired by what the author knew of Sligo, as quoted in J. W. Foster, 'Nature and Nation in the Nineteenth Century', in Foster and Chesney, *Nature in Ireland*, pp. 412–13. But contrast this with the analysis offered in H. O'Connell, *Ireland and the Fiction of Improvement* (Oxford, 2006), esp. the Introduction.

92 O'Connell, *Ireland and the Fiction of Improvement*, pp. 8–9.

93 Blackbourn, *Conquest of Nature*, p. 67.

94 See ibid., pp. 67, 70 on the mixed understandings of conquest of nature.

95 See, for example, J. Adelman and F. Ludlow, 'The Past, Present and Future of Environmental History in Ireland', *Proceedings of the Royal Irish Academy: Archaeology, Culture, History, Literature*, cxiv (2014), pp. 359–91, esp. pp. 378–9; Neeson, 'Woodland in History and Culture', pp. 140–1.

consolidating, power. And if all environmental struggles are at root 'a struggle among interests about power', as Donald Hughes has suggested, then perhaps Palmerston (and others) were just doing what Cicero had encouraged when he suggested that 'Finally by means of our hands we endeavour to create as it were a second world within the world of nature'.[96] Or, rather than a world within nature, perhaps a blend of landscape and human histories. This is what scholars such as Paul Sutter and William Cronon have pointed to in articulating the importance of seeing all environments as essentially hybrid, in which human and other forces have interacted in ways that, ultimately, cannot be untangled.[97] Thus, if 'human history and culture cannot be easily isolated from environmental forces and circumstances',[98] then understanding Palmerston's management of his estates as an economic and political enterprise necessarily involves understanding it equally as an agricultural and horticultural, or environmental, one.

So when an English landlord such as Palmerston, schooled in the ways of the Scottish Enlightenment, employed techniques and expertise that were not only Irish but also Scottish and French, was this then an exercise in making an Irish landscape more European? Consciously, or not, was this about remaking it as a more-modern, that is, enlightened, project? If so, it might, perhaps, reframe the way in which we read his comment, noted earlier, that:

> England might be a great & powerful & prosperous Country without Ireland; Ireland never could be a considerable state without England: This is not a Question of Patriotism or national spirit; it is a Question of Geography & Statistics.

96 D. Hughes, *What is Environmental History?* (London, 2006), p. 24.

97 P. S. Sutter, 'The World with Us: The State of American Environmental History', *Journal of American History*, c (2013), p. 96; W. Cronon, 'Introduction: In Search of Nature', in W. Cronon, ed., *Uncommon Ground: Toward Reinventing Nature* (New York, 1995), pp. 23–56.

98 Sutter, 'The World with Us', p. 96.

# 3

# 'A Voice for Ireland': Isaac Butt, Environmental Justice, and the Dilemmas of the Irish Land Question

*Colin W. Reid*

Irish environmental history is a curiosity. It is, formally speaking, a component of Irish historiography still in its infancy.[1] Yet, a case can be made that a huge segment of nineteenth-century Irish historiography is—unwittingly—environmental history. Much of this historiography concerns agrarian distress, landlord–tenant relations, and successive campaigns for land reform; in other words, the political, social, and economic use of the natural environment. For all this focus on land—and it was, of course, *the* political and social question of the day for much of the nineteenth century in Ireland—the general reluctance of historians to conceptualize the Irish experience in terms of the history of the environment is striking. The land question dominated political debate in Ireland (and, indeed, in Britain), and it is this aspect which has shaped the historiography for decades. Yet, while contemporary arguments for and against land reform were often couched in the language of political economy and legal rights, it was, for advocates of change, also a question of environmental justice.

Recent theories of environmental justice tend to stress the importance of indigenous rights and local control over lands and natural resources in the face of external state or capital interest.[2] In many respects advocates of land

1 J. Adelman and F. Ludlow, 'The Past, Present and Future of Environmental History in Ireland', *Proceedings of the Royal Irish Academy: Archaeology, Culture, History, Literature*, cxiv (2014), p. 360.

2 See, for example, L. Westra, *Environmental Justice and the Rights of Indigenous Peoples: International and Domestic Legal Perspectives* (London, 2008). For an arresting case study, see K. Jacoby, *Crimes against Nature: Squatters, Poachers, Thieves, and the Hidden History of American Conservation* (Berkeley, CA, 2001).

reform in nineteenth-century Ireland also used languages that chime with notions of 'indigenous' rights and the struggle for local control over natural resources, opening up a Pandora's box of national and sectarian conflict. To demand 'the land of Ireland for the people of Ireland', as was common during the Land War (1879–82), implied that it was clear who constituted the people of Ireland and, as such, who was entitled to make use of Ireland's natural resources. That this was far from clear in practice, as many historians have shown, did not diminish the political traction notions of indigeneity came to have. Indeed, the turn to Home Rule from 1870, and especially after 1879, manifested an upsurge in nationalist sentiment in combination with frustration that long-standing environmental grievances had not been addressed. Nationalists imagined that a Dublin parliament might legislate for agrarian fairness where Westminster had failed to.

Despite this development, advocates of environmental justice in nineteenth-century Ireland tended to eschew absolute claims based on indigeneity and instead sought to create a fair contractual relationship between landlord and tenant that could resolve the tensions immediate to Ireland and create the conditions that would allow the peoples of Ireland to benefit from membership of the Union and the British Empire. It was commonly believed that the extreme poverty and economic hardship that blighted many parts of the Irish countryside in the decades following the Act of Union of 1800 was as a by-product of a land tenure system that denied farmers and cultivators legal rights to their holdings.[3] 'Justice' was explicitly at the heart of the land agenda. For example, in 1866, *The Nation* appealed for legislation to 'place the relationship between landlord and tenant on the basis of justice',[4] whereas John Stuart Mill argued that land reform was not a trojan horse for undermining the legal claims of landowners, but the realization of 'justice' for the 'actual cultivators of the soil', who have been denied the right of security of tenure and fair rents.[5]. For such thinkers, achieving environmental justice in nineteenth-century Ireland was a question of reforming rather than overcoming the structures of imperial power.

When viewed in this way, Ireland's natural environment becomes central to the struggle over political rights during the nineteenth century. If we consider the Irish land question as a political construct, it provides an arresting case study of the politics that underpin what scholars have depicted as the 'hybridity' of the natural environment. This contends that

3 K. Theodore Hoppen, *Governing Hibernia: British Politicians and Ireland, 1800–1921* (Oxford, 2016), p. 119.

4 *The Nation*, 26 May 1866.

5 John Stuart Mill, *Chapters and Speeches on the Irish Land Question* (London, 1870), p. 85.

all environments are a fusion of human activity and nature.[6] In sum, if land ownership in Ireland was maintained by the legislative, judicial, and coercive apparatus of the state, which inhibited the development of a class of small, profitable farmers fundamental to the success of the agrarian economy, so the agrarian environment should be regarded as produced by a partnership between the state, landlordism, and nature, which the tenantry had little capacity or incentive to effect.

A close engagement with Isaac Butt's writings on the land question demonstrates that his grasp of this dynamic was central to his advocacy of land reform and, by extension, environmental justice. The unproductive condition of Ireland's natural environment, Butt recognized, which disadvantaged all the peoples of Ireland but the tenantry in particular, was produced and sustained by injustice. As one of the leading advocates for reform, Butt articulated a justice-based argument that transcended the narrow confines of political economy (although, as we shall see, the 'dismal science' had a prominent role to play in Butt's thought); this perspective made the ability or otherwise of the British state to deliver justice through land legislation the issue par excellence in judging the success of the Act of Union settlement. For Butt, the case for environmental justice, therefore, was nothing less than a testing ground for the legitimacy of the British constitution in Ireland. If Britain passed this test, the conditions for the material transformation of the Irish environment could be created, with all the benefits this would deliver for its peoples.

## I

Isaac Butt was one of the most significant figures of his day: as a pioneer of political economy, a man of letters, a gifted lawyer, and the father of the Home Rule movement, his contribution to intellectual and political life was considerable. He occupied almost a unique position within the Irish political spectrum, combining unionist and federalist tendencies within a broader patriotic Tory framework that offered, in stark contrast to later unionists, only *conditional* support for the Union between Britain and Ireland.[7] The logic of his position was that Union had to benefit Ireland, and not just the imperial state. The experience of the nineteenth century suggested otherwise, and it

6 P. S. Sutter, 'The World with Us: The State of American Environmental History', *Journal of American History*, c (2013), p. 96.

7 C. W. Reid, '"An Experiment in Constructive Unionism": Isaac Butt, Home Rule and Federalist Political Thought during the 1870s', *English Historical Review*, cxxix (2014), pp. 332–61 and J. Spence, 'Isaac Butt, Irish Nationality and the Conditional Defence of the Union, 1833–70', in D. George Boyce and A. O'Day, eds., *Defenders of the Union: A Survey of British and Irish Unionism Since 1801* (London, 2001), pp. 65–89.

was this realization that pushed Butt towards an embrace of federalist ideas to reposition the British connection on the basis of equality. Crucially, it was the land question which accelerated this shift. Butt's perspective on land shaped his ideas about the nature of representative government and the duties of the state to its citizens. On becoming the Whately Professor of Political Economy at Trinity College Dublin in 1836, the land question became central to Butt's evolving vision for a Union that was sensitive to conditions in Ireland and responsive to Irish ideas. He continued to write about land during the Great Famine in the 1840s and the three decades that followed. These writings shed light on what 'land', both in the abstract and physical world, meant to one of nineteenth-century Ireland's most engaging intellectuals and the wider nationally minded Tory *milieu* from which he came.

Butt was born in Glenfin, County Donegal in September 1813. He was the son of Robert Butt, a Church of Ireland clergyman, and his second wife, Berkeley Cox, who claimed descent from Bishop Berkeley. When Isaac was a year old, the family moved to Stranorlar, a small town in the Finn Valley of Donegal. His early life was spent in the shadow of the Bluestack mountains. Butt later reflected that the imposing scenery which formed the backdrop to his childhood forged his character: 'his moods ... changed from vivid enthusiasm to blackest melancholy, like the clouds sweeping across the face of the hills'.[8] At the age of fifteen, Butt left Donegal and, like his father before him, entered Trinity College Dublin. During his undergraduate career, two major pieces of electoral reform passed through Westminster: Catholic Emancipation in 1829 and the Reform Act of 1832. On his graduation, then, Butt, who was from staunch Protestant and conservative stock, found an irreversibly transformed Irish political world from that of his childhood.

In response, and along with a handful of talented young men associated with Trinity College Dublin, Butt founded the *Dublin University Magazine* (*DUM*) in 1833. The *DUM* was the intellectual journal for Irish Tories fearful of a new, increasingly democratic age, and sought instead to recast the idea of Irish Protestant national leadership within the Union. Under Butt's editorship of the *DUM* (1834–8), the journal explicitly rejected the O'Connellite Repeal movement's monopoly on Irish national identity, and the notion that the political problems in Ireland can be solved in London without reference to Irish 'facts'. The *DUM* proved as critical of the unresponsiveness of the British state in Ireland as it was of the emergence of an exclusively Catholic Irish nationalism.[9] Butt's later standing as the founding father of the Home Rule

8 T. de Vere White, *The Road of Excess* (Dublin, 1945), p. 4.

9 On the *Dublin University Magazine* in general, see J. Spence, 'The Philosophy of Irish Toryism, 1833–52: A Study of Reactions to Liberal Reformism in Ireland in the Generation between the First Reform Act and the Famine, with Especial Reference

movement in the 1870s has been interpreted as the final stage of a conversion from his firebrand Orange unionism to nationalism, but this is incorrect.[10] It is more accurate to see Butt's championing of federal Home Rule as the logical conclusion to his lifelong hostility to Catholic nationalist ideas of quasi-independent self-government and his disdain for an overly centralized Union.[11] Butt lamented the asymmetry of the nineteenth-century constitution: Britain prospered and took pride in the maturity of its statehood, while Ireland was impoverished and beset with sectarian divisions.[12] Ireland's problems stemmed from an aloof governing elite that largely ignored Irish concerns, veering instead from failed attempts at assimilation to heavy-handed coercion, thereby undermining the spirit of union.[13] As Butt aggressively put it in 1867, 'The whole government of Ireland, from the Battle of the Boyne to the present day, is one long-continued exception of Ireland, not only from British law, but from the British constitution'.[14] Nowhere was that most obvious than the land question.

A large part of Butt's political writings was dedicated to the land question and, as such, they entered the Irish political canon.[15] His ideological opponents recognized his significance also: Michael Davitt included a chapter on Butt in his *Fall of Feudalism in Ireland*, which depicted Butt as a bridge between the mid-nineteenth-century reformist Tenant Right League and the more notionally radical Land League. Butt's writings on land from the 1860s, Davitt claimed, became 'text-books for Land League speakers and writers'.[16] Butt's influence on Land League thinking, particularly those members who embraced the 'three Fs'—fair rent, fixity of tenure, and free sale—is sometimes overlooked. The process of 'deButtification' within the Home Rule party once Charles Stewart Parnell ascended to the leadership ensured that Butt's political ideas, particularly federalism, were, for a time, consigned to the nationalist dustbin of history. While federalism was a victim of the Parnellite takeover of the Home Rule movement, Butt's thinking on the land question enjoyed considerably more posthumous success.

to Expressions of National Feeling among the Protestant Ascendancy' (University of London Ph.D. thesis, 1991).

10 This approach is adopted in D. Thornley, *Isaac Butt and Home Rule* (London, 1964).

11 Reid, 'An Experiment in Constructive Unionism', pp. 337–8.

12 Isaac Butt, *Home Government for Ireland: Irish Federalism! Its Meaning, its Objects, and its Hopes* (4th edn., Dublin: Irish Home Rule League, 1874), p. 42.

13 Hoppen, *Governing Hibernia*, p. 2.

14 Isaac Butt, *The Irish People and the Irish Land: A Letter to Lord Lifford* (Dublin, 1867), p. 183.

15 For example, see Charles Read and T. P. O'Connor, *The Cabinet of Irish Literature: Selections from the Work of the Chief Poets, Orators, and Prose Writers of Ireland* (4 vols., London, 1878–80), iv. pp. 165–7.

16 M. Davitt, *The Fall of Feudalism in Ireland: Or the Story of the Land League Revolution* (London, 1904), p. 81.

## II

Butt's writings on land tended to avoid aesthetic reflection. He published a novel in 1840 called *Irish Life: In the Castle, the Courts, and the Country*, a pessimistic take on the uncertainties of the Ireland during his early life. Despite its title, the book had little to say about 'the Country', dwelling more on the Whig and Repeal indiscretions within Dublin society.[17] Butt's historical novel from 1848, *The Gap of Barnesmore*, also lacked any reflection on the natural environment, despite its geographically suggestive title.[18] Butt's literary imagination was shaped primarily by the inheritance of Irish history, namely an antagonistic form of denominational politics, and his view of land was shaped by its utility for the security and happiness of the greatest number of people. 'I care nothing for the abstract prosperity of mountains, and villages, and plains—I care a good deal for the comfort and happiness of the people whom God has now placed on them'.[19] Environmental justice was, for Butt, rooted in a humanist conception of nature, largely devoid of overt ecological or topological influence. Given the popularity of travel writings about Ireland during the nineteenth century, with many contemporaneous texts celebrating the picturesque landscapes across the island, this may seem surprising. But this is to ignore the essential humanist underpinnings of this tourist material. While such books were nominally about the scenery in Ireland, the real subject in many cases was in fact the Irish people and, in particular, their relationship to the 'Irish Question'. Butt was not unique in conceptualizing the land question as one of flesh and bone rather than crops and soil.

Butt was operating within an environmental rubric that emphasized the need for justice from the perspective of those excluded from legal rights over the land. The importance of this humanist approach was encapsulated by the title of one of his most important tracts, *The Irish People and the Irish Land* (1867): the people, not accidentally, came before the land. Land was, in Butt's estimation, a question of political economy and, as he got older, increasingly a marker to judge the value of the Union settlement of 1801. At the age of twenty-three, Butt was elected to the Whately Professorship in Political Economy at Trinity College Dublin, a newly established chair to promote the understanding of the new 'science' in Ireland.[20] Part of the remit of the post

17 [Isaac Butt], *Irish Life: In the Castle, the Courts, and the Country* (3 vols., London, 1840); Spence, 'The Philosophy of Irish Toryism', pp. 353–64; R. F. Foster, *Words Alone: Yeats and His Inheritances* (Oxford, 2011), pp. 61–3.

18 Isaac Butt, *The Gap of Barnesmore: A Tale of the Irish Highlands, and the Revolution of 1688* (3 vols., London, 1848).

19 Butt, *The Irish People and the Irish Land*, p. 180.

20 On the Whately Professorship, see T. A. Boylan and T. Foley, *Political Economy and*

was to publish lectures to promote public awareness of political economy; Butt published what he entitled an 'introductory lecture' on production in 1837, which serves as an insight into his early thinking, especially in relation to Ireland's natural environment.

Butt disliked the term 'political economy', believing that it was an unnecessarily off-putting and vague.[21] His 'immediate duty' to his audience and later readers was to illuminate 'certain truths, connected with the social condition of man'.[22] He used the example of purchasing a pair of cotton stocking, asking his audience to consider the variety of labour activity—from growing the cotton, transporting, and storing the materials to manufacturing the final product—that underpinned production.[23] This was a microcosm of a functioning economy as a whole, which was constructed on a network of often unseen labour, all working together as a machine. Political economy, for Butt, was the study of the laws which regulated this machine, particularly in the 'production, distribution and consumption of wealth'.[24] He defined 'wealth' as 'all things material or immaterial which are useful or agreeable to man' which can be transferred from one party to another. Land was identified as one such example.[25] Implicit in Butt's argument was a belief that the study of political economy shone a light on how laws *should* operate. This did not necessarily imply that the analytical frameworks established within political economy were necessarily partisan; Butt appealed to economists to ignore the 'influence of particular forms of government' and avoid the distorting prism of 'the disturbing and exciting influence of general politics'.[26] Such a plea was, however, impossible to maintain in nineteenth-century Ireland, not least for Butt himself.

Butt's approach to political economy was, like that of the later Irish economist John Kells Ingram, notable for its attempt to inject humanitarianism into the 'dismal science' in an age of seemingly unfeeling laissez-faire.[27] In rejecting the neoclassical model of economic individualism, Butt emphasized the need for the state to place dignity for its citizens at the

*Colonial Ireland: The Propagation and Ideological Function of Economic Discourse in the Nineteenth Century* ([1992] London, 2005), pp. 17–42.

21 On Butt's approach to political economy, see A. O'Day, 'Isaac Butt and Neglected Political Economists', in N. Thompson and N. Allington, eds., *English, Irish and Subversives among the Dismal Scientists* (Bingley, 2010), pp. 375–401.

22 The lecture is reproduced in T. A. Boylan and T. Foley, eds., *Irish Political Economy* (4 vols., London, 2003), i. pp. 27–56 (quote at i. p. 28).

23 Ibid., i. pp. 35–7.

24 Ibid., i. p. 37.

25 Ibid., i. p. 41.

26 Ibid., i. p. 42.

27 Richard T. Ely, 'Introduction', in John Kells Ingram, *A History of Political Economy* ([1880] London, 1923), p. xiii.

heart of its legislative programme. He was sceptical of the legislative thought that underpinned what became the Irish Poor Law Act of 1838, as it made the 'fatal mistake' of assuming that Irish poverty was 'the accident of individuals, instead of considering it as the essential and general condition of a class'.[28] While the drafters of the parliamentary bill 'prepared a plan for dotting the country with workhouses', the 'vital question'—the structural reasons why 'a large proportion of our fellow-beings [go] without sufficient food'—were not addressed. The 'condition of the poor of Ireland' Butt argued, was a disgrace, 'not only to the British empire, but to humanity'.[29] Butt sounded a radical note in alleviating the 'condition of the poor, arguing that such a plan should 'raise the condition of an entire class'. 'We want, in a word', he argued, 'not relief for occasional want, but a remedy for general and permanent destitution'.[30] For Butt, this began with revising how land was used in Ireland. Taking only the existing cultivated lands, food shortages were an inevitable feature of Irish life, as the needs of the population outstripped the supply of the land. Butt urged the state, and his Irish readers, to broaden their conception of the agrarian economy to include uncultivated lands, the scale of which in Ireland (Butt estimated there were some 5,000,000 acres of waste) more than made up the food deficiency. Unemployed labourers should be given the task of cultivating waste. Thus, Butt offered what he believed was a practical alternative to the poor law, one which dealt with the causes of destitution rather than the consequences.[31]

The logic of Butt's stance on cultivating Ireland's unused terrain was that the 'possession of land' was 'the first necessary of life'.[32] In a rural society such as Ireland, competition for land was acute, which contributed to the scenes of desperation in the countryside. More land, then, had to be sourced, to ease destitution, with the creation of larger farms to replace the 'miserable holdings' that blighted the country.[33] In reimagining the rural economy, Butt desired to reduce Irish emigration by ensuring that farming became more efficient and profitable. Given the nature of economic life in Ireland, land was a matter of life and death; proposed action (or inaction) by the state regarding agricultural reform demanded serious scrutiny. The extension of the poor law into Ireland, in Butt's reading, barely scratched the surface of economic hardship, which rested on profound inequalities and wastages on the land.

Butt's hostile reception to the poor law can be traced to what the *DUM* denounced in 1835 as the British tendency to govern Ireland according to

28 Isaac Butt, *The Poor-Law for Ireland Examined* (London, 1837), p. 8.
29 Ibid., p. 5.
30 Ibid., p. 11.
31 Ibid., p. 19.
32 Ibid., p. 30.
33 Ibid., p. 31.

'English theories', with a disregard for 'Irish facts'.[34] In a lecture at Trinity College Dublin in 1840, Butt openly preached the merits of protective duties for Irish industries as a means to encourage economic growth 'in reference to the circumstances of *our* own country'.[35] The hiberno-centric remit of the lecture can be read as a critique of the insensitive nature of British rule, coupled with a rejection of the free trade consensus that seemingly profited advanced industrial nations at the expense of small, rural countries. 'Let us have the artisans and the peasantry of our own country comfortable and prosperous', he asserted, 'before we weigh the chances of feeding those of the world at large'.[36] Butt presented protectionism, then a discredited idea within the intellectual traditions of transatlantic political economy (especially in the aftermath of the repeal of the Corn Laws in 1846), as the only humane economic policy to pursue in Ireland. 'It is vain to speak of the blessings of increasing national wealth', he declared, 'if to this you sacrifice the comforts of the poor'.[37] He avoided pressing a more doctrinal point by claiming that the binary extremes which depicted free trade as 'good' and protectionism as 'evil' were unhelpful, as the choice of economic model should rest pragmatically on the circumstances of 'each particular instance'.[38] Butt's advocacy of protectionism was such that R. D. C. Black positioned him as one of the chief 'dissenting' voices within nineteenth-century political economy.[39] In championing protectionism, Butt asserted the rural distinctiveness of the Irish economy, which differed fundamentally from the more intensified and industrial British model. And while Butt did not intend protectionism to become subsumed into identity politics, the discourse of exceptionalism he helped to create during the 1840s fed into advances in nationalist thinking. In 1847, against the backdrop of the Famine, John Mitchel published a fiery polemical called *Irish Political Economy*. Mitchel injected Anglophobia into Butt's basic protectionist premise, with dramatic effect: 'if Irishmen are hungry, it is that Englishmen may be filled—if Irish tradesmen are idle, it is that English ones may be fully employed, at good wages—if Irish labourers perish, it is simply that English labourers may flourish'.[40]

34 'English Theories and Irish Facts', *Dublin University Magazine*, vi (1835), pp. 682–96.
35 Isaac Butt, *Protection to Home Industry: Some Cases of its Advantages Considered* (Dublin, 1846), p. 25 (emphasis added).
36 Ibid., p. 75.
37 Ibid., p. 61.
38 Ibid., p. 84. Cf. Butt's interpretation with the influential argument that free trade was adopted to morally restrain the economy in B. Hilton, *The Age of Atonement: The Influence of Evangelicalism on Social and Economic Thought, 1785–1865* (Oxford, 1992), pp. 36–70.
39 R. D. C. Black, 'The Irish Dissenters and Nineteenth-Century Political Economy', in A. E. Murphy, ed., *Economists and the Irish Economy* (Dublin, 1984), pp. 120–37.
40 John Mitchel, ed., *Irish Political Economy* (Dublin, 1847), p. iv.

By this point, with the onset of the Famine, the blurring between the 'political' and the 'economy' in Butt's world view became more pronounced, as he moved closer to the contemporaneous Young Ireland critique of the Union.[41] His strongly held humanist interpretation of environmental justice was central in this process.

As a political economist, Butt was interested in the structural inequalities in Ireland, and how best they should be remedied. With the onset of the Famine, he began more forcefully to connect the wretched nature of the Irish economy to the centralized design of the Union. This marked a definite shift in tone. Butt's reputation in the Ireland of the early 1840s, after all, was forged in part by his celebrated speech against Daniel O'Connell's motion in favour of Repeal of the Union at the Dublin Corporation, an address considered as one the most coherent defences of Irish unionism ever made.[42] In 1847, Butt published *A Voice for Ireland*, which, because it came from a unionist perspective, should be considered one of the most incendiary texts of the period. While nominally a sustained critique of the devastating consequences of the application of free trade principles in Ireland only one year after the repeal of the Corn Laws, Butt's emergence as the 'voice for Ireland' saw him move politically closer to his radical Young Ireland contemporaries. 'Can we wonder if the Irish people believe', Butt fumed, '*and believe it they do*—that the lives of those who have perished, and who will perish, have been sacrificed by a deliberate compact to the gains of English merchants'. The humanitarian failure of theories of political economy had, Butt asserted, 'created among all classes a feeling of deep dissatisfaction, not only with the ministry but with English rule'.[43] The Famine, in other words, grimly brought to the surface the tensions inherent not merely within competing ideas of political economy, but the Union itself. In face of the catastrophe, Butt questioned the ability of the Westminster parliament to represent Ireland, if it continued to prioritize 'English' concerns:

> If the Union be not a mockery, there exists no such thing as an English treasury. The exchequer is the exchequer of the United Kingdom ... Ireland has been deprived, by the Union with England, of all separate power of action ... Irishmen were told, indeed, that in consenting to a Union which would make them partners with a great and opulent nation, like England, they would have all the advantages that might be expected to flow from such a Union. How are these expectations to be

41 See R. P. Davis, *The Young Ireland Movement* (Dublin, 1988).

42 Isaac Butt, *A Speech Delivered in the Corporation of Dublin, on the 28th February 1843, on Mr O'Connell's Motion to Petition for a Repeal of the Legislative Union* (Dublin, 1843); White, *Road of Excess*, p. 67.

43 Isaac Butt, *A Voice for Ireland: The Famine in the Land* (Dublin, 1847), p. 10. The pamphlet was originally published in the *DUM*.

realized, how are these pledges to be fulfilled, if the partnership is only to be one of loss, and never of profit to us?'[44]

The logic of such a stance was the restoration of the Irish parliament, an idea with which Butt flirted until the botched Young Ireland rebellion of 1848.[45]

*A Voice for Ireland* advocated a string of initiatives to alleviate the ruination of the Famine, from investment in infrastructure (particularly railways) to state-sponsored emigration. Butt repeated his earlier call for the cultivation of wastelands, suggesting that without Ireland's full agrarian potential being realized, the country could not support a population of eight million people. To feed the entire population, Butt lamented that radical change in the use of land was needed before the onset of crisis in the 1840s:

> We should have had the accumulated products of the industry of many former years—factories in which the savings of the present would enable the owners to advance food and sustenance to thousands of workmen, who now are wretched squatters on her bogs or her fields; a soil highly improved, the result of a long and careful tillage, farmsteads provided with all the appliances of high and successful cultivation, and all the means of large expenditure within the reach of the farmer; we should, in a word, have in the hands of all the different classes of society what political economists called capital ... Were Ireland such a country as this, it could support, from its natural resources, far more than its present population.[46]

But Ireland was not such a country, and, for this, the government was to blame. The calamity of the Famine was, in Butt's estimation, 'an imperial one', and the cost for relieving destitution should be borne not by Ireland itself, but the wider empire. Failure to apply 'a liberal and generous policy' amidst such grave scenes of human tragedy threatened to confirm within the Irish mind that the London parliament had 'abdicated the functions of government for Ireland'. The logical outcome was a 'separate legislative existence' for Ireland.[47] Butt concluded *A Voice for Ireland* by emphasizing his conservative instincts. If Ireland is ruined through Famine, he warned, the first class to fall will be the landed gentry. Such an eventuality promised to unleash 'the horrors of Ireland's social revolution'.[48]

44 Butt, *A Voice for Ireland*, p. 21.
45 See C. W. Reid, 'Constitutional Rhetoric as Legal Defence: Irish Lawyers and the Languages of Political Dissent in 1848', in K. Hughes and D. M. MacRaild, eds., *Crime, Violence and the Irish in the Nineteenth Century* (Liverpool, 2017), pp. 120–3.
46 Butt, *A Voice for Ireland*, p. 43.
47 Ibid., pp. 53–4.
48 Ibid., pp. 55–6.

This was an important strand of Famine-era political argument. While rooted in imperialism and hostility to the Repeal movement's fusion of Catholicism and democracy, Butt had no hesitation in firing a warning shot across the bow of the British state in advocating a radical constitutional rethink of the nature of the United Kingdom. While Butt's interventions were driven by a humanitarian impulse, underpinning his argument was the belief that the core problem that blighted Irish life was an inability to reform the rural economy. For Butt, land reform—or, more to the point, the lack of land reform—crystallized doubts he harboured about the nature of the legislative union. 'Land', as a question of environmental justice, became enmeshed in a broader problem concerning representative government. Butt's 'conversion' to federalism in 1870 was not a sudden volte-face; the seeds of his later 'turn' to Home Rule were planted much earlier.[49]

## III

For much of the 1850s and 1860s, Butt sat in parliament, where he was one of the most active Irish members.[50] His rise to national prominence during the 1860s occurred against the backdrop of a renewed shadow of famine, with appalling weather inflicting catastrophic damage to the agrarian economy from 1859 to 1864,[51] as well as the formation of the Fenian movement, the transatlantic revolutionary society dedicated to the overthrow of the British state in Ireland. The grim possibility of a rebellion by the Fenians hung in the air during 1865; revolt became reality in Kerry, Cork, Dublin, and Limerick in 1867, although these outbreaks were quickly suppressed. The Famine and Fenian threats triggered a new wave of pamphlet and other print materials setting out potential solutions to the resurrected 'Irish question'. Perhaps the most famous was John Stuart Mill's angry denunciation of Ireland's experience of Union, *England and Ireland*, which interpreted the Fenian rebellion not as an expression of doctrinaire republicanism, but a manifestation of social—and particularly agrarian—grievance.[52] Indeed, there was something in this: the Fenian Proclamation of the Irish Republic from 1867 contained a line condemning the landed disfranchisement of the 'real owners of the soil', 'THE IRISH PEOPLE'.[53]

49 Reid, 'An Experiment in Constructive Unionism', p. 340.
50 O'Day, 'Isaac Butt', p. 394.
51 J. S. Donnolly, 'The Irish Agricultural Depression of 1859–64', *Irish Economic and Social History*, iii (1976), p. 36.
52 John Stuart Mill, *England and Ireland* ([1868] 5th edn., London, 1869), p. 6.
53 *Freeman's Journal*, 8 March 1867.

By the time of the rebellion of 1867, Butt had re-emerged as one of the leading advocates of land reform in Ireland, acting as the chief spokesmen for tenant rights for small farmers.[54] Two of his major works from the period, *Land Tenure in Ireland: A Plea for the Celtic Race* (1866) and *The Irish People and the Irish Land* (1867), gained much traction in Ireland.[55] In 1870, Butt published *The Handbook of the Land*, which brought together excerpts from his previous works. An additional publication, *The Irish Querist* (1867), borrowed the stylistic approach of Bishop Berkeley's *The Querist* (1735–7) in asking a plethora of provocative questions, the vast majority of which centred on land.

Butt's writings from the late-1860s developed some strands of thought that had been laid down previously. The connection between the dire economic state of Ireland and the nature of centralized government appeared again, with a proto-Home Rule stance adopted. Butt's land pamphlets were particularly concerned with the need for fixity of tenure on land holdings, one of the key 'three Fs' that the post-Famine Tenant League championed. Insecurity of tenure was, Butt argued in 1866, the evil that underpinned everything that was wrong with the Irish land system. He painted a bleak picture of an Irish peasantry denied legal rights to the land they occupied, with the fear of unfair eviction hanging over small farms across the countryside, resulting in a wretched existence for the tenants and a chronically underdeveloped rural economy. Like Mill, he believed that Fenianism grew out of an agrarian rather than a national antagonism.[56] The only solution was fixity of tenure: Butt proposed a sixty-year contract (this was extended to sixty-three in *The Irish People and Irish Land* the following year) to embed farmers firmly in their holdings.[57] This would have occurred had Ireland enjoyed functioning representative government, Butt reasoned, as 'laws and institutions are gradually moulded to the wants and wishes of the people' in nations that controlled their own political destiny.[58] In *The Irish People and the Irish Land*, Butt directly appealed for a reconciliation between 'law' and 'justice'.[59] While the 'English constitution' had been venerated in Walter Bagehot's book of the same name, which was also published in 1867, Butt re-emphasized Ireland's detachment from the rights and liberties enshrined in law, which left tenant farmers vulnerable. 'Ireland is denied the real benefit of English justice by the

54 Steele, *Irish Land and British Politics*, p. 54.

55 Both pamphlets were widely cited in contemporaneous reform-minded literature. The *Freeman's Journal* also serialized extracts from *Land Tenure in Ireland* on 7, 9, and 16 July 1866.

56 Isaac Butt, *Land Tenure in Ireland: A Plea for the Celtic Race* (3rd edn., Dublin, 1866), p. 9.

57 Ibid., p. v.

58 Ibid., p. 71.

59 Butt, *Irish People and the Irish Land*, p. 6.

fiction that we are one country with England'.[60] The withholding of justice on the land not only damaged the rural economy; the very legitimacy of the British state in Ireland was impaired. The immediate origins of Butt's shift to Home Rule can thus be found in his reflections on the land question.

Michael Davitt depicted Butt as 'moderate precursor' of 'Irish revolutionary land reform'—in other words, the Land League—while James Fintan Lalor was its 'prophet'.[61] Certainly, Butt's demand for fixity of tenure appears 'moderate' when compared with Lalor, who advocated land nationalization, and the later agrarian activism spearheaded by Davitt. Butt himself presented his argument not as a radical rethinking of the land system, but as a 'conservative measure' which reconciled 'the rights of property with the right of the people to live upon their land'.[62] By 1870, John Stuart Mill had similarly moved towards fixity of tenure, describing such a measure in parliament as a 'modest demand'.[63] In keeping with his cautious approach, Butt also urged the British political classes to concede piecemeal reform of the Irish land system now, or risk facing a much more radical demand for change in the future.[64] It is wrong, however, to view Butt solely through the prism of 'moderation'. Taken on Butt's word in 1870, federal Home Rule was a conservative and moderate measure. The leading constitutional lawyer of the day, A. V. Dicey, believed, however, that Butt's federalism motion was the most radical scheme ever put before the British parliament.[65] Similarly, while Butt's advocacy of tenant right might appear cautious, fixity of tenure involved the legislative weakening of the long-established property rights of landlords in Ireland, and was thus condemned by the landed interest as revolutionary. As a point of comparison, Mill, the champion of the same 'modest' scheme of reform, was dubbed a 'rabble-rouser' and supporter of 'dangerous "isms"', such as 'Irish nationalism, anarchism, and communism', for his support for fixity of tenure.[66] Terms such as 'conservative', 'moderate', and 'radical' were in the eye of the beholder when it came to the question of Irish land. Butt's writings on land were steeped in national sentiment, but were, at heart, an appeal for a more equitable and just union.

The radicalism or otherwise of the proposal of fixity of tenure aside, what is tangible is the extraordinarily angry language Butt used to discuss the

60 Walter Bagehot, *The English Constitution* (London, 1867); Butt, *Irish People and the Irish Land*, p. 268.
61 Davitt, *Fall of Feudalism*, p. 79.
62 Butt, *Land Tenure*, p. 92.
63 Mill, *Chapters and Speeches on the Irish Land Question*, p. 106.
64 Ibid., p. 104.
65 Reid, 'An Experiment in Constructive Unionism', p. 359.
66 D. Winch, *Wealth and Life: Essays on the Intellectual History of Political Economy in Britain, 1848–1914* (Cambridge, 2009), p. 81.

general state of Ireland during the 1860s. While his Famine-era writings were tinged with a frustration borne from grief, culminating in his judgement that the union represented a partnership of loss to Ireland, Butt's later writings deployed language more akin to his revolutionary contemporaries, James Fintan Lalor and John Mitchel. Mitchel's *The Last Conquest of Ireland (Perhaps)* was first published in 1861, and may have re-popularized contemporaneous usage of the term 'conquest' to describe the relationship between Britain and Ireland, and the political and social defects in the latter. Certainly, Butt was not immune from this trend. 'Conquest' was not used in *A Voice for Ireland* from 1846 but peppered his pamphlets from the 1860s. Butt overtly framed the unjust circumstances of the contemporary land question as an extension of the centuries-old domination of Ireland. 'The whole system of landed property', Butt proclaimed, 'is regarded by the great mass of the people as an alien institution, all its rights are looked upon as enforced by conquest, and maintained only by a foreign force'.[67] Similarly, he described a 'war' between the owners and occupiers of the land.[68] In *The Irish People and the Irish Land*, which was a direct reply to criticism Butt received from the landed interest, the conquest theme was developed further. 'The perpetual origin of misery and degradation', he suggested, 'has been the fact that the great mass of the people have been treated as belonging to a conquered race'.[69] The arbitrary power of eviction in Ireland was sanctioned by 'the sword of Oliver Cromwell'.[70] The 'old monster grievance of the Irish race' was landlordism, 'which embodies the policy and the passions of territorial conquest'.[71] The 'whole system of landed property in Ireland', Butt lamented in *The Irish Querist*, was 'the creation of English conquest'.[72]

Despite its relatively modest aims, the presentation of Butt's case was, then, steeped in a much more emotive discourse than he had hitherto deployed regarding land or political economy. This was perhaps the fusion of two impulses: first, growing frustration with the condition of Ireland within the United Kingdom and empire; second, a deliberate appeal to, and channelling of, Mitchelite sentiment. While Butt's instincts remained conservative and imperial-minded from the 1840s to the 1870s, there were definite shifts in how he conceived Ireland's future role within the United Kingdom, and these were articulated in a language that increasingly emphasized Ireland's lack of legal, political, and economic parity with England. The 'conquest'

67 Butt, *Land Tenure in Ireland*, p. 42.
68 Ibid., p. 20.
69 Butt, *Irish People and the Irish Land*, p. 91.
70 Ibid., p. 246.
71 Ibid., p. 118.
72 Isaac Butt, *The Irish Querist: A Series of Questions Proposed for the Consideration of All Who Desire to Solve the Problem of Ireland's Social Condition* (Dublin, 1867), p. 31.

metaphor was thus a rhetorical tool used by Butt to forcibly articulate his discontent. At the same time, Butt was perhaps utilizing a truculent language to convince radicals in Ireland to follow the path of constitutional action. His admiration after 1867 for the patriotic virtue, if not the methods, of the Fenian rebels was well known, most notably by his legal defence of several prisoners.[73] Several years before the rebellion of 1867, Butt confided in the journalist A. M. Sullivan that the danger was not a successful Fenian rising, but 'an abortive attempt', which invited ridicule and discredited 'every appeal to our physical strength'.[74] His published works can thus be interpreted as an alternative to Fenianism (and its potential failures), a political approach which appealed to patriotic sentiment, but funnelled through constitutional and decidedly non-violent means.

One of the ironies of this was that Butt, at times, 'out-Mitcheled' his radical contemporaries. Despite his chosen title, Mitchel only used the word 'conquest' several times in passing in *The Last Conquest* and did not develop a sophisticated analysis of what that term meant in an Irish context. Butt deployed the word much more frequently than Mitchel and linked it explicitly to the injustices in the Irish land system. Such powerful language contributed to Butt's rise as the leading figure of tenant right during the late 1860s. He mirrored, in style if not in substance, the revolutionary writings of Lalor from the 1840s, which aimed to fuse the causes of democratic self-government and land nationalization in a bid to 'repeal the Conquest'.[75] The overlap in tone between Lalor and Butt illuminates the fiery nature of the latter's approach. Lalor believed that the 'conquest' was 'still in existence'—a sentiment echoed by Butt some twenty years later—as the 'landlord holds his lands by right and title of Conquest, and uses his powers as only a conqueror may'.[76] Similarly, Lalor described the claim of the people of Ireland to the soil as a 'natural right'; Butt later appealed to the 'supreme power in the state to interfere against law and property' in Ireland, 'and in favour of natural justice and right'.[77] Butt frequently deployed a slavery trope in his writings about land, describing the Irish farmer as a 'slave' unable to escape from legally sanctioned selfdom, in contrast to the English farmer who is a 'freeman' because he enjoyed legislative protections and a stake in representative government.[78] Lalor described the 'whole community' in Ireland

73 Also see Isaac Butt, *Ireland's Appeal for Amnesty: A Letter to the Right Hon. W. E. Gladstone, MP* (Glasgow and London, 1870), p. 15.

74 Isaac Butt to A. M. Sullivan, 6 September 1865, Isaac Butt Papers, National Library of Ireland [NLI] MS 831/unnumbered letters.

75 J. F. Lalor, *'The Faith of a Felon' and Other Writings*, ed. M. Ramón (Dublin, 2012), p. 22.

76 Ibid., p. 137.

77 Ibid., p. 26; Butt, *Land Tenure in Ireland*, p. 81.

78 See, for example, Butt, *Land Tenure in Ireland*, pp. 89, 102 and Butt, *Irish People and the Irish Land*, pp. 9, 24, 69, 178, 235, 267, 280.

as being 'made up of tyrants, slaves, and slave-drivers'.[79] Lines by Butt, such as 'There never will be peace in Ireland while it [feudal domination] exists', echo the militant tone of Lalor's *Irish Felon* newspaper.[80]

The form of Butt's rhetoric on land during the 1860s was deliberately saturated with pathos, offering a window into his personal radicalization in the age of Fenianism, as well as the wider societal attitudes which he aimed dually to shape and reflect. Davitt later observed that there were two sides of Butt: he 'frequently put the lawyer and constitutionalist in the background in his speeches and writings, and spoke as the descendant of a Rapparee enemy of landlordism and as a Donegal witness of the horror and infamy of the social crime of eviction'.[81] Indeed, in *The Irish Querist*, Butt raised the spectre of the Glenveagh evictions, one of the most notorious mass dislodgement of tenants during the nineteenth century, which occurred in County Donegal in 1860. The most shocking aspect of the Glenveagh evictions for Butt was that when the landlord, John George Adair, gave the order to remove dozens of families from his estate, it was enforced by 'a large body of the Queen's troops'.[82] Butt's anger, however, arose because such an act of cruelty was not confined to one Irish locality or another; if mass evictions did not occur at Glenveagh, they would inevitably happen somewhere else, as the entire system was morally bankrupt. For Butt, this was an issue beyond the locale. Glenveagh demonstrated that the state was complicit in the monstrously unjust actions of landlords, which poisoned the constitutional tie between Ireland and Britain. As Butt potently suggested, 'landlordism and the land system was the impersonation of English domination and misrule'; but the state, in his reading, had the power to emancipate itself from the tangled web of Irish history and transform Ireland's feudal serfs into freemen.[83] Land reform, then, for Butt, represented nothing less than the trigger for Irish improvement and the transformation of the Union into a genuine partnership of equals.

## IV

Butt died two years before the Land Act of 1881 was passed, which introduced the security to tenant farmers for which he had long campaigned. Writing at that time, Henry George presented the Irish land problem as a binary clash: 'Either the land of Ireland rightfully belongs to the Irish landlords, or

79 Lalor, '*Faith of a Felon*', p. 104.
80 Butt, *Irish People and the Irish Land*, p. 20.
81 Davitt, *Fall of Feudalism*, p. 91.
82 Butt, *Irish Querist*, p. 10.
83 Butt, *Land Tenure in Ireland*, p. 9; *Irish People and the Irish Land*, p. 235.

it rightfully belongs to the Irish people; there can be no middle ground'.[84] There was, of course, a 'middle ground', which Butt represented. His appeal for fixity of tenure was permeated with a radical critique of a land system that was linguistically similar to the writings of James Fintan Lalor, but offered a much more modest (if still radical) endpoint than Lalor's advocacy of nationalization. Despite Butt's depiction of landlordism in Ireland as the personification of conquest and English misrule, he was, as Philip Bull has noted, giving the landed class an opportunity to rescue itself 'from the brink of disaster'.[85] Fixity of tenure was presented by Butt as the means 'to retain nearly all of proprietary right, and yet make the people free'.[86] Like Lalor, Butt was calling for the repeal of the conquest; but crucially he was not calling for the destruction of landlordism. Indeed, Butt firmly believed that the gentry families of Ireland had an opportunity to resume a national leadership role once the antagonism of agrarian injustice was removed. Buttite political thought suggested that the Act of Union had ruined the landed class of Ireland, as they had 'unlearned national politics' in their embrace of English interests over Irish concerns.[87] Butt recognized that land reform presented the gentry with a chance to redeem itself nationally, should they be brave enough to embrace it.

Butt thus distinguished between landlords and landlordism: while the latter was a rotten institution in urgent need of reform, he continued to extol the potential leadership virtues of the former. By 1870, Butt had gravitated towards a scheme to federalize the United Kingdom, thus granting Ireland its own parliament.[88] He was at pains to present the case for federalism in conservative terms, with his pamphlet on the issue peppered with references to an Irish parliament acting as a bulwark to nationalism and radicalism. Essential to this was Butt's conception of a bicameral parliament, with an Irish House of Lords granting the gentry an opportunity 'to place themselves in the front of a national government'.[89] Butt frequently connected the injustices of the land system to a lack of representative government: once 'Irish ideas' permeated the legislature and executive—whether that be at Westminster or an Irish parliament—the amelioration of Ireland would swiftly follow.[90] While Butt hypothesized that the improvement of tenant farmers in Ireland required Irish influence over the levers of legislative power, he was, perhaps curiously, eager to keep the issues of 'land' and 'self-government' separate from

84 George, *Land Question*, p. 38.
85 Bull, *Land, Politics and Nationalism*, p. 65.
86 Butt, *Irish People and the Irish Land*, p. 297.
87 F. H. O'Donnell, *A History of the Irish Parliamentary Party* (2 vols., London, 1910), i. p. 59.
88 Reid, 'An Experiment in Constructive Unionism', pp. 332–61.
89 Butt, *Home Government*, p. 56.
90 Butt, *A Voice for Ireland*, p. 54; *Land Tenure*, p. 71.

one another once the Home Rule movement came into existence. In 1870, the first resolution adopted by the Home Government Association of Ireland was that the purpose of the organization is to achieve 'a national parliament'; the second resolution affirmed that this was the *only* object of the movement.[91] The Association was envisaged by Butt as a one-issue pressure group, or, as James Golden has dubbed it, 'a 'discussion society'.[92] Land was not, however, an issue to be discussed.

The first incarnation of a Home Rule society contained elements from across the entire spectrum in Ireland, from the aristocratic interest to Fenians, Tories to old Repealers. The need to avoid damaging the fragile common ground between these groups was paramount. Butt was all too aware of the hostility of the landed gentry to his ideal of fixity of tenure, with Lord Lifford condemning such a scheme as 'communistic' and Lord Dufferin believing tenant security was a Trojan horse to destroy landlordism.[93] To protect the uneasy alliance between the gentry and agrarian reformers under the Home Rule banner, land was thus not on the official agenda of the Home Government Association. While Butt had emerged as one of the leading lights of land reform and self-government by 1870, he was anxious not to fuse the two issues. The campaign for self-government required a broad support base, drawing on all socio-economic, political, and denominational groups; rooting this national vision in agrarian politics was self-defeating because it narrowed the potential support base. The irony of this stand came only later, with the might of Parnellism resting on the ability of Butt's successor as chief of the Home Rule movement, Charles Stewart Parnell, in fusing the national demand for Home Rule and agrarian agitation.[94] In many ways, Parnellism was Butt's unintended creation: he gave the Parnellite movement a vocabulary aggressively to press the case for land reform and the intellectual rationale for self-government. It was the blurring of land and national agitation which made the Irish Party of the 1880s so formidable.

When he was the leader of the Home Rule movement, Butt complained that 'the Irish people are sadly deficient in the practical knowledge of politics. They are looking out for something sensational which can hardly even be given to them without wish of mischief'.[95] In an imagined dialogue between the

91 The resolutions are printed in A. M. Sullivan, *New Ireland: Political Sketches and Personal Reminiscences of Thirty Years of Irish Public Life* (16th edn., London, 1877), pp. 344–5.

92 Butt, *Home Government*, p. vi; J. J. Golden, 'The Protestant Influence on the Origins of Irish Home Rule, 1861–1871', *English Historical Review*, cxxviii (2013), p. 1514.

93 Quoted in Butt, *Irish People and the Irish Land*, 6; Lord Dufferin, *Mr Mill's Plan for the Pacification of Ireland Examined* (London, 1868), p. 7.

94 P. Bew, *Enigma: A New Life of Charles Stewart Parnell* (Dublin, 2011), pp. 73–7.

95 Butt to Father O'Dwyer, n.d. [1873], Isaac Butt Papers, NLI MS 831/unnumbered letters.

ghosts of Butt and Daniel O'Connell written by Henry Traill in 1884, Butt's timidity is mocked by the old Repeal chief, and is presented as the reason for the stagnant performance of the Home Rule cause in parliament during the 1870s.[96] Such jibes were written in the context of a renewed nationalist agitation partly catalysed by Parnell's leadership, by which standard Butt's 'weak' leadership paled in comparison. This generalization of Butt overlooks a number of vital points, chiefly the radical nature of his advocacy of fixity of tenure and, later, Home Rule, two policies which challenged absolute property rights and the nature of the British constitution respectively. It also obscures the extent to which Butt created and perpetuated political languages to promote substantial agrarian and constitutional reform, and the sophistication of his conservative radicalism. The ambition, vision, and hope in Butt's writings were immense.

Butt's understanding of 'land' was informed by political and economic considerations, but at heart he was driven by a sense of environmental justice, centring on the occupants of the soil. Indeed, by the 1860s, Butt believed that justice, politics, and economic thought were pushing in the same direction, making change irresistible:

> On the question of Land Tenure in Ireland, I maintain that precisely the same result follows from the dictates of justice, which oblige us to give the Irish people the means to live; from the suggestions of political expediency, which point out how mischievous is the dissatisfaction of a whole population; and from the lessons of political economy, which tell us how the most can be made of the resources of the land.[97]

Butt also imagined land reform as the key to unlocking Irish potential, the means by which landlords could make common cause with tenant farmers, leading to an overall political, social, and economic improvement in the condition of Ireland. The failure of the British state to legislate for meaningful land reform in Ireland—or, as Butt frequently put it during the 1860s, the ongoing perpetuation of the conquest—contributed to his reconsideration of Ireland's constitutional relationship with the United Kingdom. The natural environment in Ireland was thus a fiercely contested political question which profoundly shaped wider constitutional thought.

96 H. D. Traill, *The New Lucian: Being a Series of Dialogues of the Dead* (London, 1884), pp. 136–60.
97 Butt, *Land Tenure*, p. 66.

# 4

# 'In The Open Country': Nature and the Environment during the 'Monster' Meeting Campaign of 1843

*Huston Gilmore*

On 23 May 1997, in a commemorative issue to mark the 150th anniversary of the death of Daniel O'Connell, the *Kerryman* newspaper contained the following description of the 'monster' meeting campaign of 1843: 'O'Connell held monster meetings in about thirty centres during the summer and autumn of 1843. Between 3 and 4 million people attended'.[1] To the extent that O'Connell's 'monster' rallies remain in the public consciousness, these grossly inflated numbers of attendance have been the norm.[2] It has been argued that the O'Connellite campaign of repeal *événements* during the spring and summer of 1843 not only 'transformed public life in Ireland' on the eve of the Famine, but brought Ireland 'to the brink of agrarian revolution'.[3] To analyse the role the environment played during the O'Connellite 'monster' meetings it is necessary to note that there is a dearth of sources beyond contemporary newspaper coverage to illuminate this aspect of the campaign, and that hitherto scholarship focusing on nature and the environment in Ireland in this period has been overwhelmingly drawn to the catastrophe of the

1 *Kerryman*, 23 May 1997.

2 See, for example, a report of a recreation of a 'monster' meeting at Ballitore, Co. Kildare, which unquestioningly states 800,000 people had attended the 1843 Mullaghmast meeting. 'Dan still draws the crowds: Town re-enacts O'Connell event', *Irish Press*, 4 Oct. 1993. O'Connell's most recent biographer largely reproduces the inflated numbers of the repeal press throughout his account: P. M. Geoghegan, *Liberator: The Life and Death of Daniel O'Connell 1830–1847* (Dublin, 2010), pp. 133–66.

3 D. Dwan, *The Great Community: Culture and Nationalism in Ireland* (Dublin, 2008), p. 141; P. Gray, *Famine, Land and Politics: British Government and Irish Society, 1843–50* (Dublin, 1999), p. 49.

Famine (1845–51).[4] Nonetheless, it is possible to use newspaper accounts to consider how the 'monster meetings' allowed contemporaries to undertake the temporary dominance of public landscapes by associating in crowds, and the manner in which these occasions constituted a political charivari when norms could be contested, such as the deployment of decorative flora as a political symbol in defiance of then-existing property rights: landscapes are integral to modes of self-formation and identity, as well as being connected to structures of power and authority.[5] There are thus insights to be drawn from the consideration of O'Connellite 'monster' meetings as a phenomenon that occurred within a charged political atmosphere, and did so within specific environments in which place, identity, and a discourse of nationalist grievance was negotiated through a historicization of the Irish landscape.[6]

It is perhaps surprising that O'Connell did not adopt a more explicit discourse of 'improvement' during 1843, although it is in keeping with his characteristic vagueness regarding economic matters throughout his career, and it is telling that the Repeal Association did not explicitly turn towards such matters until two years after the 'monster meeting' campaign when the members of 'Young Ireland' had attained a greater prominence in the Repeal Association.[7] Nonetheless, for O'Connell, the achievement of legislative independence was paramount, a catholicon from which all the solutions to Ireland's problems would necessarily follow; it was a *sine qua non* that the Act of Union had retarded the socio-economic potential of Ireland, which he held to be 'one of the most abundantly fertile countries on the globe'.[8] His argument for repealing the Union was thus couched with explicit references to the natural world, and his stated desire was to reclaim 'the superior fertility, station, and natural qualities' of Ireland that he claimed had withered under rule from Westminster.[9] Hence, O'Connell employed a discourse of nature

4 See, for example, Juliana Adelman and Francis Ludlow's exploratory survey, 'The Past, Present and Future of Environmental History in Ireland of Environmental History in Ireland', *Proceedings of the Royal Irish Academy. Section C: Archaeology, Celtic Studies, History, Linguistics, Literature*, cxiv (2014), pp. 359–91, which begins its consideration of the modern period with the Famine. In J. W. Foster's seminal *Nature in Ireland*, political matters in the 1840s receive only passing mention in reference to subsistence crises: D. Synott, 'Botany in Ireland', in J. W. Foster, ed., *Nature in Ireland: A Scientific and Cultural History* (Dublin, 1997), pp. 171–2.

5 J. Malpas, 'Place and the Problem of Landscape', in J. Malpas, ed., *The Place of Landscape: Concepts, Contexts, Studies* (London, 2011), p. 14.

6 G. Ó Tuathaigh, 'Irish Land Questions in the State of the Union', in F. Campbell and T. Varley, eds., *Land Questions in Modern Ireland* (Manchester, 2016), pp. 4–5.

7 K. Theodore Hoppen, 'Riding a Tiger: Daniel O'Connell, Reform, and Popular Politics in Ireland, 1800–1847', *Proceedings of the British Academy*, c (1999), p. 141.

8 D. O'Connell, *A Memoir of Ireland, Native and Saxon* (Dublin, 1843), p. 45.

9 D. O'Connell, *A Full and Revised Report of the Three Days Discussion in the Corporation of Dublin on the Repeal of the Union* (Dublin, 1843), p. 11.

and nation drawn from a moral universe in which, notwithstanding the endemic nature of Irish poverty, he was:

> struggling for a soil fruitful in every virtue, abundant in its productiveness. No summer sun scorched their plains into barrenness—no winter snow ever chills them into sterility. They are green, fertile, and lovely ... and with a people boasting of the highest class of moral and intellectual qualities, the day of Ireland's liberty was approaching—the Repeal was at hand.[10]

O'Connell was fiercely proud of his remote estate in Derrynane, made repeated reference to the natural beauty surrounding it throughout his career, and pronounced himself to be not 'a ferocious demagogue ... [but] ... in truth, a gentle lover of Nature'.[11] For O'Connell, 'repeal' was a panacea, based on the Panglossian assumption that the natural predisposition of Ireland was towards prosperity, if only the curse of the Union could be lifted:

> And for what a land are we struggling? Oh, it is a lovely, a beauteous land. I am fond of the picturesque beauties of nature ... Yes, her mountains are majestic, her plains are fertile, her valleys are green and luxuriant, and sheltered; her mountain streams pour down a water power capable of turning the machinery of all the manufactures of the world ... Yes, the natural beauties and the fertility of Ireland place her foremost among the nations of the earth.[12]

To an extent, O'Connell's repeated invocation of nature may be seen as a trope deployed by a professional orator, as when at Athlone, where W. J. O'Neill Daunt mordantly recorded O'Connell imploring his audience to 'look around ... on the majestic mountains' in what he described as 'the centre of the flattest part of Ireland' (Daunt noted that 'compliance with which request would have severely tested the optics of his audience').[13] Nonetheless, O'Connell certainly took inspiration from Ireland's natural beauty to make his case for repeal, telling a Tipperary audience in April 1843 that:

> He felt there was no exaggeration in saying ... when he thought of the teeming fertility of the multitudinous fields he had recently looked

10 *Freeman's Journal*, 20 May 1843.
11 D. O'Connell to W. S. Landor, 4 October 1838, in W. J. Fitzpatrick, ed., *The Correspondence of Daniel O'Connell, The Liberator* (2 vols., London, 1888), ii. p. 152.
12 *The Nation*, 25 March 1843.
13 Diary of W. J. O'Neill Daunt, 27 June 1843, NLI MS 3040.

upon—the greenness of her hills—the fertility of her vallies [*sic*]—the verdure of her plains, and the productive power that nature and nature's God had given her ... Oh, it was a land to fill me with patriotism.[14]

The audiences O'Connell addressed, of course, acted both within and upon the environment, but the crowds that attended the 'monster' meetings in 1843 have not, until recent years, been well served by historians.[15] Undoubtedly there has been little tendency to locate a progression from what George Rudé termed the transition from 'older forms of popular movement to the new' in the crowd activity and sociability of Irish peasants that has, for some, been found in the mass platform of Chartism in Britain.[16] However, John Plotz's suggestion that Chartist crowds be understood as a 'form of language, whose effect cannot be compared to the merely paper logic of newspapers' can be equally applied to O'Connell's mass rallies.[17] The work of Gary Owens has cast the 'monster' meetings as 'elaborate secular rituals' and as 'liminal phenomena in which familiar constructs of time, space, work, and social relationships were momentarily set aside'.[18] These participatory congregations were held with the intention of instilling new perceptions of communal solidarity amongst those who attended them, as well as to project numerical strength to a wider Irish, British, and international audience.[19] In her study of protest, space, and place in industrial northern England in the half century that followed the French revolution, Katrina Navickas has suggested not only that political protests and crowd activity in this period should not be delineated as strictly 'urban' or 'rural' due to the permeable boundaries

14 *Freeman's Journal*, 20 April 1843.

15 'Human history has always and will always unfold within a larger biological and physical context'. See J. R. McNeill, 'Observations on the Nature and Culture of Environmental History', *History and Theory*, xlii (Dec. 2003), p. 6.

16 G. Rudé, *The Crowd in History: A Study of Popular Disturbances in France and England, 1730–1848* ([1964] London, 2005), p. 179. On the emergence of sociability in rural Irish life in the late eighteenth and early nineteenth centuries, see T. Bartlett, *The Fall and Rise of the Irish Nation: The Catholic Question, 1690–1830* (Savage, MD, 1992), chapter 14 (pp. 310–19).

17 J. Plotz, *The Crowd: British Literature and Public Politics* (Los Angeles, CA, 2000), p. 129.

18 G. Owens, 'Nationalism without Words: Symbolism and Ritual Behaviour in the Repeal "Monster" Meetings of 1843–45', in J. S. Donnelly Jr and K. A. Miller, eds., *Irish Popular Culture, 1650–1850* (Dublin, 1998), pp. 242–69. See also Owens's 'Constructing the Repeal Spectacle: Monster Meetings and People Power in Pre-Famine Ireland', in M. R. O'Connell, ed., *People Power: Proceedings of the Third Annual Daniel O'Connell Workshop* (Dublin, 1993), pp. 80–93.

19 J. F. Dean, *All Dressed Up: Modern Irish Historical Pageantry* (New York, 2014), p. 9. On the cross-channel and transatlantic nature of the intended audience of repeal demonstrations, see C. T. McMahon, *The Global Dimensions of Irish Identity: Race, Nation and the Popular Press, 1840–1880* (Chapel Hill, NC, 2015), pp. 11–17.

between the two, but also that mass political rallies in this period 'had similar rituals and forms ... each site of meeting's environment, topography, size, and transportation connections were different ... The materiality of the landscape shaped the location and experience of meetings'.[20] The same might be said of the O'Connellite 'monster' meeting, which was undertaken with an established form, utilized popular rituals, and which was characterized by meticulously organized political discipline within particular and changing topographies.[21] The making of a repeal 'place' was undertaken with reference to both Ireland's pre-reformation history and the natural environment (and the conflation of the two). The parish priest of Baltinglass, County Wicklow, the Revd Daniel Lawler, made this explicit:

> Look at the crumbled old churches of our land, and they will tell you the history of the past—they have survived the tempest, and still stand, like the fragments of lost history, to give the slight sketch of former terrors; the green ivy that surrounds them is a true image of the inseparable connection between the hierarchy and the people; it mantles their lofty turrets in the proud days of their glory; and now clings to them with equal fidelity when their shivered fragments lie upon the ground in scattered and forgotten ruins.[22]

All but three of the 'monster' meetings were held in or near strategically chosen country towns and involved processions from small town to rural environments, invariably with O'Connell's carriage taking centre stage—a form of what Navickas has termed 'politics on the move'.[23] O'Connellite rallies combined popular demonstrations of support with a dominant presence of the Catholic clergy: or, as O'Connell's eccentric lieutenant Tom Steele described it, 'the Liberator's progress ... has been one uninterrupted scene of popular triumph, shoutings [*sic*], bonfires—illuminations—bands—flambeaux—*not forgetting blessings*'.[24] The location of these meetings reflects the political geography of O'Connellite politics, always strongest in the provinces of Leinster and Munster.[25] Nineteen meetings, almost half

20 K. Navickas, *Protest and the Politics of Space and Place, 1789–1848* (Manchester, 2016), pp. 225, 242.
21 H. O'Connell, *Ireland and the Fiction of Improvement* (Oxford, 2006), chapter 4 (pp. 161–3).
22 Speech of the Revd D. Lawler, PP Baltinglass, *Freeman's Journal*, 8 Aug. 1843.
23 Navickas, *Protest and the Politics of Space and Place*, p. 177.
24 T. Steele to W. J. O'Neill Daunt, 17 Sept. 1843. O'Neill Daunt papers, NLI MS 10507 (18) (emphasis added).
25 K. Theodore Hoppen, *Elections, Politics and Society in Ireland, 1832–1885* (Oxford, 1984), chapter 6 (pp. 479–80).

of the total, were held in Leinster; eleven in Munster; eight in Connaught, and three in Ulster.[26] It has often been stated that no meetings were held in the northern part of the island, for fear of provoking sectarian violence, and it is certainly true that O'Connell was loath to pursue agitation in Belfast, no doubt recalling his disastrous trip in 1841 that had ended in a riot and his escape in disguise.[27] However, O'Connell's close ally W. J. O'Neill Daunt did attend meetings at Inishowen, County Donegal, on 6 August, in the company of the Catholic Bishop of Derry, and at Clontibret, County Monaghan, nine days later, while O'Connell appeared at a meeting at Carrickmacross, County Monaghan, in April.[28] Given the potential for outbreaks of violence, it is perhaps unsurprising that no 'monster' meeting was held in the north-east of Ulster, and when O'Connell did travel to Monaghan, the meeting was chaired symbolically by a former member of the Orange order, the landlord Captain Thomas Seaver, of Heath Hall, Newry.[29]

From the perspective of the historian, contemporary accounts are frustratingly vague as to the nature of the land on which the meetings were held, more often than not simply reporting that they occurred in 'fields' or 'meadows' on the outskirts of towns, without recording who owned these fields, or whether they took place on various types of common land.[30] As the campaign gained momentum and audience numbers grew, recourse was had on a number of occasions to staging them on racecourses—such as at the Curragh, on 7 May; at Croker's Hill racecourse, three miles outside of Kilkenny, on 8 June; at Ballycorey, just outside Ennis, on 15 June; and at Roscommon, where the procession moved from the 'Race-park, about two miles from the town', to 'an extensive meadow, about a mile from town, on the Ballygar road', on 20 August.[31] On other occasions, the gatherings took place on land adjacent to Catholic churches, or, in rare instances, belonging to

26 In 1844, on more than one occasion, O'Connell referenced that there had been forty-one 'monster' meetings during the campaign. See *Freeman's Journal*, 30 Jan. 1844, 30 Mar. 1844.

27 P. Maume, Preface to *The Repealer Repulsed* (Belfast, 1841; repr. Dublin, 2003), p. xix.

28 *The Nation*, 19 Aug. 1843; Journal of W. J. O'Neill Daunt, 15 Aug. 1843, NLI MS 3040.

29 *Drogheda Argus*, 29 April 1843. This is not to say such large-scale demonstrations did not occur in Ulster, but rather that they were a feature of Protestant, rather than Catholic, culture. See D. Curran, 'The Great Protestant Meeting of Dungannon, 1834', in W. Sheehan and M. Cronin, eds., *Riotous Assemblies: Rebels, Riots and Revolts in Ireland* (Cork, 2011), pp. 96–119.

30 For an overview of common land in Ireland, see J. H. Andrews, 'The Struggle for Ireland's Public Commons', in P. O'Flanagan, P. Ferguson, and K. Whelan, eds., *Rural Ireland 1600–1900: Modernisation and Change* (Cork, 1987), pp. 2–5.

31 *Freeman's Journal*, 9 May 1843; 10 June 1843; 22 Aug. 1843.

sympathetic landlords or their tenants, such as the meeting at Lismore, which was held on a deer park on Lord Devonshire's estates.[32] The *Cork Examiner* explained that this was 'an enclosure of some thirteen hundred acres which was kindly granted for the occasion, by the tenant of the lands, an honest and respectable farmer named Richard Farrell'.[33] In at least one instance, it was noted that permission to use the land on which the rally was held was granted by a Protestant (a Mr Creighton, at Clifden) but that this fact was given prominence by the speakers suggests it was a departure from the norm. Not only did Creighton give permission for the use of land 'contiguous to the bridge and waterfall', but he was also reported as having offered to 'open his mill-dams in order to allow as much of the water as possible to pass without falling over the cascade, so as to diminish the sound that might otherwise interfere with the voice of the speakers'.[34] Meetings were also staged on fair grounds, as at Donnybrook and Castlebar on 3 and 31 July respectively, or communal hillsides, such as at Tomduff, within a mile of Enniscorthy.[35] This was directly opposite Vinegar Hill, the site of a battle in which United Irish insurgents had been defeated by the British army in 1798, but which was reported as 'no longer presenting a scene of strife and contention, but one of patriotic and peaceful exertion'.[36]

Much as it would be tempting to locate a line of succession in the O'Connellite use of public land to previous periods, it instead appears that in most instances the reason for selecting a place of meeting was the pragmatic need to accommodate a pavilion, a crowd, and ease of access for O'Connell's carriage, rather than associations with a distant Gaelic past.[37] However, on a number of occasions, sites for these meetings were chosen on the basis of their historical resonances of past conflict (such as Mullaghmast, which was both the site of a victorious engagement of the United Irishmen during 1798 and a massacre of Catholic gentry in 1577), landmark Irish victories of the distant

32 See, for example, the meeting at Kells on 22 April 1843, where the platform 'was erected under the shade of a majestic sycamore tree in front of the chapel house' (*Freeman's Journal*, 25 Apr. 1843) and the meeting at Athlone on 18 June, where the meeting was held in a 'meadow' in front of the house in which O'Connell had stayed the night before, belonging to Edward Murphy, Esq. *Freeman's Journal*, 21 June 1843. See also Jim Lenehan, *Politics and Society in Athlone, 1830–1885: A Rotten Borough* (Dublin, 1999), pp. 31–2.

33 *Cork Examiner*, 25 Sept. 1843.

34 *Freeman's Journal*, 20 Sept. 1843.

35 *Freeman's Journal*, 4 July 1843, 2 Aug. 1843.

36 *Freeman's Journal*, 22 July 1843. The *Freeman's Journal* published the most extensive coverage of the 'monster' meetings in the repeal press, dispatching a special reporter to the demonstrations from the outset.

37 E. Fitzpatrick, *Royal Inauguration in Gaelic Ireland, c.1100–1600: A Cultural Landscape Study* (Woodbridge, 2004), pp. 223–5.

4.1 The 'Monster' Meeting at Tara, *Illustrated London News*, 26 August 1843

past (Clontibret, Clontarf, site of ancient battles involving Hugh O'Neill and Brian Boru respectively), or as symbols of ancient sovereignty (Cashel and the Hill of Tara).[38] O'Connell, who published his only historical work in January 1843, the verbose *A Memoir of Ireland: Native and Saxon*, was attuned to such connotations.[39] At the Rock of Cashel, on 23 May, the parish priest of Clonmel, Dr Burke, made a point of linking repeal with pre-reformation (and pre-conquest) Christianity, when he gestured to the ruins during his speech and said: 'There stood before them a venerable monument of the ancient glory of Ireland—once the dwelling of sanctity and learning, and the residence of sainted bishops and paternal monarchs. It ... should infuse into the hearts of every Irishman a desire to see those glories revived'.[40] At Tara (Figure 4.1), O'Connell bestowed a spurious degree of democratic accountability on ancient Irish kingship, when he drew attention to the meeting being held 'on the spot where the monarchs of Ireland were elected ... emphatically the spot

38 M. Cronin, '"Of One Mind?": O'Connellite Crowds in the 1830s and 1840s', in P. Jupp and E. Magennis, eds., *Crowds in Ireland, c.1720–1920* (London, 2000), pp. 139–72.
39 D. O'Connell, *A Memoir of Ireland, Native and Saxon* (Dublin, 1843).
40 *Freeman's Journal*, 25 May 1843.

from which is emanated the social power—the legal authority—the right to dominion over the further extremities of Ireland'.[41]

By 1843, authorities had been monitoring outdoor O'Connellite rallies for some years. A typical police report from 1840 noted that:

> a Repeal Meeting was held at Maynooth on the 27th Instant, at which Mr. Daniel O'Connell M.P. and one of his sons attended; it was held in the field opposite to the Roman Catholic Chappel [*sic*]. There may have been present about six thousand persons, consisting almost exclusively of small farmers and labourers, scarcely any of the middle class and less of the gentry of the county.[42]

However, it was not until the aftermath of a debate on Repeal of the Union in the Corporation of Dublin, and O'Connell's decision to devote himself solely to extra-parliamentary agitation, that the first 'monster' meeting was held at Trim, County Meath, on 19 March 1843 (Figure 4.2).[43] The Young Ireland newspaper *The Nation* claimed that 20,000 persons had attended, including 'hundreds of the comfortable yeoman and wealthy farmers', and emphasized the 'mighty stimulus given to the Repeal cause by the recent celebrated debate in the corporation'.[44] The Resident Magistrate for Meath, Captain George Despard, who was also in attendance, was less than impressed with the spectacle. He reported to Dublin that O'Connell had addressed 'some Roman Catholic priests and as nearly as I could calculate between 4 & 5,000 of the peasantry'. Moreover, he stated that all but one of the local Catholic gentry had eschewed the meeting, although he did imply there were signs that popular enthusiasm for the campaign was stirring: 'the peasantry evidently took more interest in the proceedings than I have yet witnessed'.[45]

Key to ensuring that such large assemblages of people could be summoned without a breach of the peace was the contemporary popularity of temperance.[46] One observer of the Tara meeting later emphasized that it was in the sociability of the occasion, rather than the oratory from the platform,

41 *Freeman's Journal*, 16 Aug. 1843.

42 Report of Sub-Inspector G. Fitzgibbon, Celbridge, Co. Kildare, 23 Dec. 1840, National Archives of Ireland, Chief Secretary of Ireland Office Registered Papers (hereafter NAI CSORP)/1840/20943 (Kildare).

43 D. Leahy, *A Review of the Principal Facts Connected with the Rise, Progress, Conclusion and Character of the Recent State Prosecutions in Ireland, by a Barrister* (London, 1845), p. 13.

44 *The Nation*, 25 April 1843.

45 G. Despard to E. Lucas, 19 Mar. 1843, NAI CSORP/1843/22/5275 (Meath).

46 On the Temperance movement in Ireland, see Elizabeth Malcolm's *Ireland Sober, Ireland Free: Drink and Temperance in Nineteenth-Century Ireland* (Dublin, 1986); C. Kerrigan, *Father Mathew and the Irish Temperance Movement* (Cork, 1992); and P. A. Townend, *Father Mathew, Temperance and Irish Identity* (Dublin, 2002).

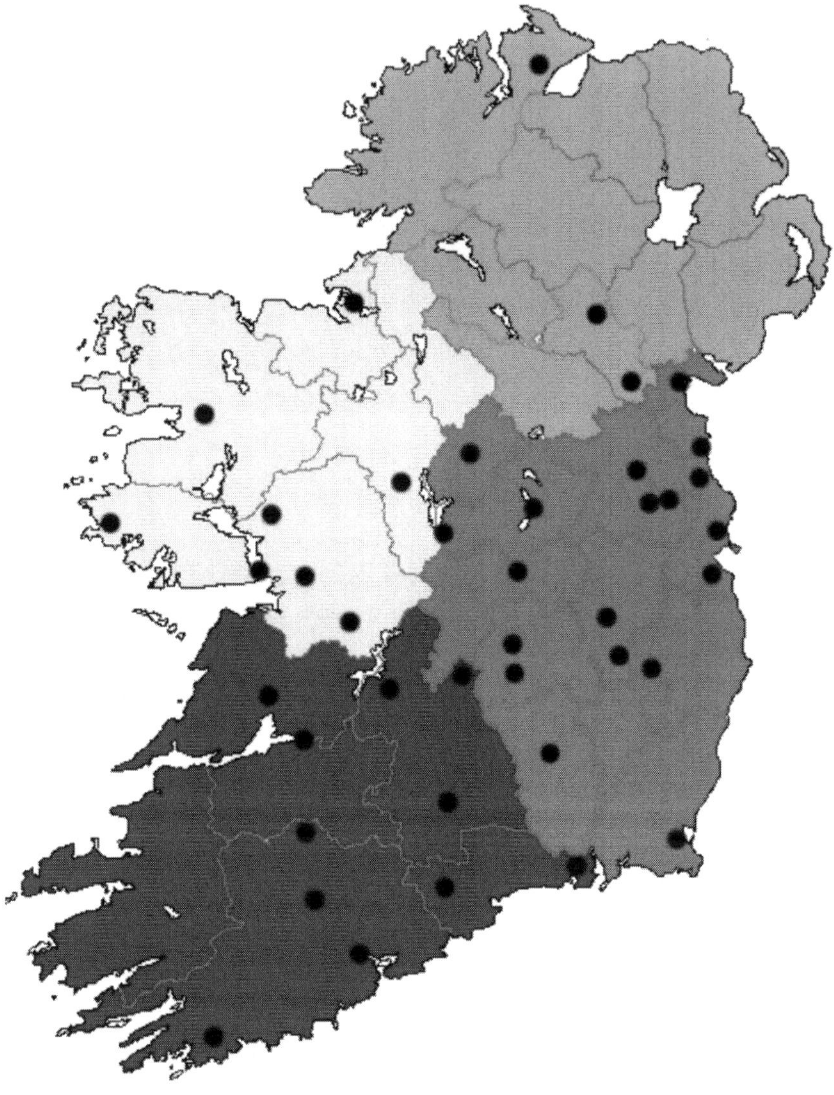

4.2 Repeal 'Monster' Meetings, 19 March 1843–1 October 1843

that most people partook, pointing to the 'holiday' aspect of proceedings: 'not a fiftieth part of the assemblage was within hearing, or even sight of the hustings. Numerous booths and tents had subsequently been erected in various directions along the crest of the hill, and also along its eastern declivity, where dancing, soda-water, ginger-beer, fruit, and different kinds

Repeal 'Monster' Meetings, 19 March 1843–1 October 1843

| | |
|---|---|
| 19 March – Trim, Co. Meath | 29 June – Dundalk, Co. Louth |
| 9 April – Bellewstown, Co. Meath | 3 July – Donnybrook, Co. Dublin |
| 18 April – Roscrea, Co. Tipperary | 9 July – Waterford, Co. Waterford |
| 19 April – Limerick, Co. Limerick | 16 July – Tullamore, King's Co. |
| 23 April – Kells, Co. Meath | 16 July – Swords, Co. Dublin |
| 25 April – Carrickmacross, Co. Monaghan | 20 July – Wexford, Co. Wexford |
| 4 May – Sligo, Co. Sligo | 23 July – Tuam, Co. Galway |
| 7 May – Curragh, Co. Kildare | 30 July – Castlebar, Co. Mayo |
| 14 May – Mullingar, Co. Westmeath | 6 August – Inishowen, Co. Donegal |
| 18 May – Charleville, Co. Cork | 6 August – Baltinglass, Co. Wicklow |
| 20 May – Cultra, Co. Galway | 15 August – Tara, Co. Meath |
| 23 May – Cashel, Co. Tipperary | 15 August – Clontibret, Co.Monaghan |
| 25 May – Nenagh, Co. Tipperary | 20 August – Roscommon, Co. Roscommon |
| 25 May – Ballinakill, Queen's Co. | 10 September – Loughrea, Co. Galway |
| 28 May – Longford, Co. Longford | 17 September – Clifden, Co. Galway |
| 5 June – Drogheda, Co. Louth | 24 September – Lismore, Co. Waterford |
| 8 June – Kilkenny, Co. Kilkenny | 1 October – Mullaghmast, Co. Kildare. |
| 11 June – Mallow, Co. Cork | |
| 15 June – Ennis, Co. Clare | |
| 18 June – Athlone, Co. Westmeath | |
| 22 June – Skibbereen, Co. Cork | |
| 25 June – Galway, Co. Galway | |

of refreshments, were relished by many thousands, instead of [the] oratory'.[47] An account of the Mullaghmast meeting stressed both the commercial and recreational nature of the event, at which an estimated 200 temporary booths were erected, and 'Tables were set out, at which refreshments of all degrees, from "mountain dew" down to tea and coffee, might be procured. In holes in the ground in front of the tents fires of turf were lighted, in which cauldrons with huge masses of beef and pork bubbled away'.[48] Similarly, a sceptical reporter at the Lismore meeting made the claim that 'The majority of those present seemed to have mustered as for recreation ... The assembled multitude appeared to have no notion of foolishly trying to hear the speeches. A *display*—a *demonstration*—seemed alone to be the object'.[49]

47 J. Johnston, *A Tour in Ireland; with Meditations and Reflections* (London, 1844), pp. 73–4.

48 An account of the Mullaghmast meeting from the *Morning Chronicle*, reprinted in the *Cork Examiner*, 24 Nov. 1843.

49 An account from the *Northern Whig*, reprinted in the *Kerry Evening Post*, 4 Oct. 1843.

O'Connell, for his part, repeatedly praised the sobriety and respectability of his audiences. Not everyone was so impressed. The German tourist Jacob Venedy, who published an account of his time in Ireland during what O'Connell had proclaimed the 'Repeal year', remarked that 'uneasiness, inattention, and rudeness characterized the multitude', noted that 'most of those on the platform were certainly not teetotallers', and recounted O'Connell raising a glass of champagne to toast the temperance leader, Father Mathew.[50] Nonetheless, Despard, when reporting on the largest of the 'monster' meetings at Tara, was struck by the fact that 'The peasantry was most peaceful, civil & good humoured, and not a drunken man to be seen'.[51] Temperance bands appear to have been ubiquitous at the meetings, often marching in quasi-military formations, and so in a general atmosphere of speechifying and cheering, we should note the prevalence of music—widely featured in 'elections and their connected happenings ... [which] ... invigorated rural and urban life'—as part of the 'monster' meeting environment.[52] Despard also implied both the performative aspect of the 'monster' meeting and highlighted the physical distances people had travelled to attend when he reported to Dublin Castle that:

> It did appear very remarkable to me that very shortly after Mr. O'Connell arrived thousands left the ground without even endeavouring to hear him speak, and I was surprised at observing that almost all the country people assembled appeared quite careless as to whether or not they heard any of the speeches ... Every county surrounding Meath and some distant ones such as Wexford, Wicklow & Longford furnished their quotas, a man told me on the ground that he came from the barony of Shelenakin [*sic*] in Wexford that 2,000 men came with him & 3,000 from Kildare joined them.[53]

Gary Owens has followed Donal MacCartney in terming these meetings 'hedge-schools' of politics, in which the masses were educated in the nationalism of Repeal.[54] However, perhaps a more apposite metaphor would be to describe them as popular politicized pilgrimages, which harnessed what Hoppen has termed 'the close of the sacred and the profane, so characteristic of late eighteenth- and early nineteenth-century Catholicism' under strict clerical and, in

50 Jacob Venedy, *Ireland and the Irish during the Repeal Year, 1843*, trans. W. B. MacCabe (Dublin, 1844), pp. 39–40.

51 G. Despard to E. Lucas, 16 Aug. 1843, NAI CSORP 22/16/583 (Meath).

52 K. Theodore Hoppen, 'Landlords, Society and Electoral Politics in Mid-Nineteenth-Century Ireland', in C. H. E. Philpin, ed., *Nationalism and Popular Protest in Ireland* ([1987] Cambridge, 2002), p. 305.

53 G. Despard to E. Lucas, 16 Aug. 1843, NAI CSORP 22/16/583 (Meath).

54 D. MacCartney, *The Dawning of Democracy: Ireland, 1800–1870* (Dublin, 1987), p. 153.

the form of the Repeal Wardens, elite lay control.[55] Certainly, some of the largest gatherings took place in symbolically sacred landscapes that saw a confluence of 'Catholicism of the chapel' with 'a popular Catholicism of the natural landscape'.[56] Nor should it be forgotten that the repeal campaign took place within a contested environment in the context of an 'ongoing attempt to transform the Catholic landscape into a reformed one', and that predominantly Catholic repeal demonstrations may be seen as a riposte to generations of Irish Protestants regularly assembling in public 'to assert their ascendancy over the papist majority in open-air ceremonies and processions'.[57] The repeal 'monster' meeting thus took place in a contested landscape with contested symbolic meanings.[58]

These events invariably entailed O'Connell's arrival by carriage in a market town, after staying at the home of a parish priest or local dignitary the night before.[59] He would then travel in procession, accompanied by the organized trades of the town, to a pre-arranged site where a pavilion had been erected (itself an example of the creation of a political environment demarcated between elite participant and spectator). After the main demonstration and a litany of orations, O'Connell proceeded to attend a ticketed public dinner in the town that evening, usually for between 200 and 600 people, in premises (more often than not, a hotel) invariably furnished with a separate gallery for the 'ladies'.[60] Here speeches were delivered, food and drink served, and loyal toasts drunk in an elite culture of politicized conviviality that naturally excluded the vast majority of those who had participated in the day's demonstration. O'Connell delivered virtually the same speech at every meeting and dinner, promising that Repeal would guarantee a panoply of measures designed to appeal to tenant farmers, such as fixity of tenure, the abolition of the tithe rent-charges, and the building of local roads and bridges (not, it has been noted, issues of primary concern to the peasantry).[61]

55 K. Theodore Hoppen, 'Riding a Tiger: Daniel O'Connell, Reform, and Popular Politics in Ireland, 1800–1847', in T. C. W. Blanning and P. Wende, eds., *Reform in Great Britain and Germany, 1750–1850* (Oxford, 1999), p. 121.

56 D. W. Miller, 'Landscape and Religious Practice: A Study of Mass Attendance in Pre-Famine Ireland', *Éire-Ireland*, xl (Spring/Summer 2005), p. 99.

57 A. Walsham, *The Reformation of the Landscape: Religion, Identity, and Memory in Early Modern Britain and Ireland* (Oxford, 2011), p. 549.

58 P. J. Duffy, *Exploring the History and Heritage of Irish Landscapes* (Dublin, 2007), p. 194.

59 It has been estimated that O'Connell's 1843 schedule entailed over 5,000 miles of travel. See O. MacDonagh, *O'Connell: The Life of Daniel O'Connell, 1775–1847* (London, 1991), p. 510.

60 On the involvement of women in O'Connellite politics, see M. O'Dowd, 'O'Connell and the Lady Patriots: Women and O'Connellite Politics, 1823–45', in A. Blackstock and E. Magennis, eds., *Politics and Political Culture in Britain and Ireland, 1750–1850: Essays in Tribute to Peter Jupp* (Belfast, 2007), pp. 283–99.

61 Cronin, '"Of One Mind?": O'Connellite Crowds in the 1830s and 1840s', p. 150.

The day's proceedings were widely reported in the repeal press, alongside details of an upsurge in the repeal 'rent'—the voluntary fund submitted to the Repeal Association—which peaked in the third week of June.[62] The 'monster' meetings thus also took place within a vibrant Anglophone print culture, and their performative aspects, as well as the highly visible role of the Catholic clergy, was undertaken with an eye to how they would be eventually reported.[63] This is not to say they occurred in a completely Anglophone environment: at Galway, the *Freeman's Journal* correspondent noted that the streets were 'very densely thronged by the country people ... talking Irish in loud accents as they went along'.[64] Despite being a native Irish speaker, O'Connell has acquired a notoriety as an undertaker of the Irish language, not least from his oft-quoted remark that he was 'too utilitarian' to regret its gradual abandonment.[65] This reputation owes much to the writings of cultural revivalists of the late nineteenth century, and to those of postcolonial theorists a century later.[66] However, O'Connell could not have had any concept of the precipitous decline of Irish in the post-Famine world, and contrary to his reputation he did, on occasion, use Irish as a political medium.[67] In particular, at a number of Repeal dinners, grace was said in Irish, explicitly linking Repeal with both Gaelic identity and the Catholic Church. Before 450 people dined in the Temperance Hall in Skibbereen, County Cork, the blessing was given—in Irish—by the Revd Matthew Horgan, parish priest of nearby Blarney (Figure 4.3). *The Nation* printed his prayer in translation:

> O Lord! Protect old Erin / And hear our cry to thee.
> We beseech thee save us / From all our enemies,
> From tithe oppression free us / Protect our Holy Faith
> O spare our Liberator / Till Erin has Repeal!

62 M. Keyes, *Funding The Nation: Money and Nationalist Politics in Nineteenth-Century Ireland* (London, 2011), p. 172.

63 H. Gilmore, '"The Shouts of Vanished Crowds": Literacy, Orality and Popular Politics in the Campaign to Repeal the Act of Union, 1840–1848', *Interdisciplinary Studies in the Long Nineteenth Century*, xviii (2014), p. 6.

64 *Freeman's Journal*, 28 June 1843.

65 J. A. Murphy, 'O'Connell and the Gaelic World', in *Daniel O'Connell: Portrait of a Radical*, p. 40. See also R. English, *Irish Freedom: The History of Nationalism in Ireland* (London, 2006), p. 138 and B. Fanning, *Histories of the Irish Future* (London, 2015), p. 234. The original anecdote, from a conversation held fifteen years before the time of writing, is in W. J. O'Neill Daunt, *Personal Recollections of the Late Daniel O'Connell* (2 vols., London, 1848), i. p. 14.

66 See, for instance, D. Hyde, 'The Necessity for De-Anglicising Ireland', in *The Revival of Irish Literature: Addresses by Sir Charles Gavan Duffy, K.C.M.G., Dr. George Sigerson, and Dr. Douglas Hyde* (London, 1894), pp. 115–61, esp. pp. 130–1; D. Kiberd, *Inventing Ireland: The Literature of the Modern Nation* (London, 1996), pp. 133, 616.

67 R. Uí Ógáin, *Immortal Dan: Daniel O'Connell in Irish Folk Tradition* (Dublin, 1995), pp. 41–3.

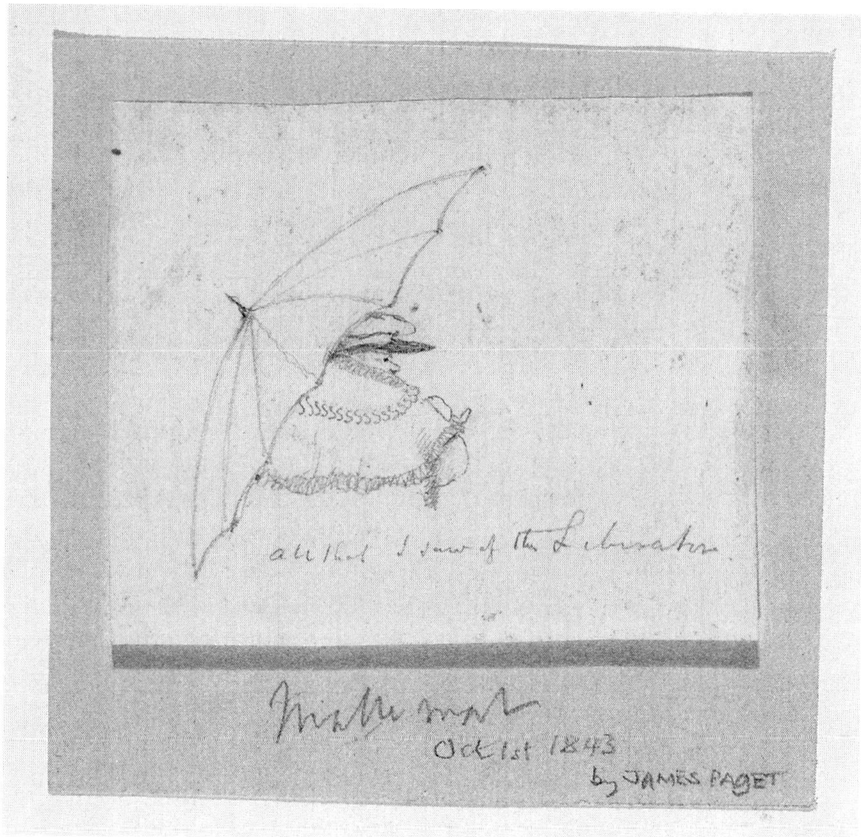


4.3 'All that I saw of the Liberator': James Paget, Mullaghmast, 1 October 1843
Image courtesy of the National Library of Ireland

Notwithstanding his historical reputation, in response to Horgan's Irish orison in 1843, O'Connell revealed a different side to his own view on the language:

> My esteemed and revered Friend Father Horgan, in that beautiful grace with which he spoke *in the proper language first*, and then badly translated into English afterwards, gave an Irish line that I confess delighted me. It meant in the vulgar Saxon tongue, 'Ireland will be herself again'.[68]

A further aspect of the 'monster' meeting that has received little attention from historians is the physical environment in which they were held, and

68 *The Nation*, 1 July 1843 (emphasis added).

which would have had an immediate bearing on those who attended: that is to say, the prevailing elements.[69] As Daunt privately noted in February 1843, 'This agitation for repeal, though not without its moments of pleasing excitement, is less than an agreeable task in piercing weather'.[70] At the first 'monster' meeting in Trim the clearing weather was held to be an augury of the successful achievement of Repeal, whereas at Bellewstown in County Meath the event was staged 'notwithstanding the extreme inclemency of the day'.[71] In May, a meeting at Nenagh in County Tipperary had to be cut short due to heavy rain.[72] At the meeting in Kilkenny, a pavilion had been erected in the town which was reported as being 'the realization of the gorgeous fancies of fairy life, and when brilliantly lit up with gas, for which an abundant supply of pipes and fittings had been laid'; nonetheless, 'The tremendous rain which fell without ceasing on Wednesday spoiled the decorations, and the high winds the succeeding night ... carried away a considerable portion of the roof'.[73] It certainly should be noted that the largest of the gatherings took place in the high summer months, when the threat of rain was diminished, and the hours of daylight the longest.[74] By 10 September, O'Connell was reported as having presented himself 'under the shelter of an umbrella', while an admittedly hostile depiction of the 'monster' meeting at Clifden, written by Benjamin Disraeli's brother Ralph, recalled that 'the rain unfortunately fell in torrents', and that O'Connell had ended the occasion 'sitting under his umbrella, amid the roar of the elements and the shouts of repealers ... [looking like] the double impersonation of Puck and Mephistopheles'.[75]

A suggestive aspect of the role of nature and the environment at these gatherings is to be found in press reports of the meetings, which emphasized that popular support was predominantly expressed by recourse to the symbolic use of foliage and greenery to denote political allegiance. At Drogheda, the *Freeman's Journal*'s correspondent reported that along the demonstration's route:

69 The weather unsurprisingly affected all types of outdoor political meetings. One correspondent from County Waterford informed the Repeal Association in April 1843 that they had organized a repeal meeting in their chapel yard for the first time, but that 'the meeting was not very numerously attended as the day was wet'. Thomas O'Neill Powers to T. M. Ray, 27 Apr. 1843, Repeal Association papers, NLI MS 13625 (1).

70 Journal of W. J. O'Neill Daunt, 10 Feb. 1843, NLI MS 3040.

71 *Freeman's Journal*, 21 Mar. 1843, *The Nation* 15 Apr. 1843.

72 27 May 1843.

73 *Freeman's Journal*, 10 June 1843.

74 D. Ó Giolláin, 'Celebrations and the Rituals of Life', in E. F. Biagini and M. E. Daly, eds., *The Cambridge Social History of Modern Ireland* (Cambridge, 2017), p. 301.

75 *Freeman's Journal*, 10 Sept. 1843; R. Disraeli, *A Visit To The Wild West: A Sketch of the Emerald Isle, Picturesque and Political, during the Past Autumn* (London, 1843), pp. 16–18.

> The cabins of the peasantry were enveloped so completely in laurel branches, that the proportions of their humble structures were in many cases concealed altogether from view (a circumstance not always to be regretted); flags, colours, and ribbons floated from various eminences, and triumphal arches, festooned with flowers and laurels, and bearing appropriate inscriptions, were thrown across the road.[76]

These carefully stage-managed processions through triumphal arches en route to the demonstrations, and ostentatious displays of pageantry, must be viewed as employing an orchestrated dramaturgy, and a deliberately constructed nationalist iconography.[77] The prevalence of the use of boughs, laurel leaves, and natural greenery at the 'monster' meetings was not just a way round the prohibition of flying the green flag of nationalist Ireland, although it was that too.[78] In part, it was a matter of the paucity of printed material in a society characterized by widespread illiteracy and rural poverty: when dispatched to cover the meetings in September, *The Times* correspondent reported from Loughrea that 'There was nothing new in the banners ... many I recognized as having done duty on former occasions, and some as having been dispatched by the same conveyance that brought me from Dublin'.[79] Nor was the use of foliage a recent innovation. On one occasion at least, the 'Head Pacificator' Tom Steele referred to the use of green boughs as symbols in political processions during the Emancipation campaign of the 1820s.[80] Moreover, Steele himself promoted this co-option of the natural for political purposes in contradistinction to British authority: at Tara, when exhorting a crowd to maintain O'Connell's peaceful methods of agitation, he wielded 'a very light green bough, the emblem of his office as O'Connell's Head Pacificator of Ireland' and announced that 'he would not exchange that little green bough ... for any regal or imperial sceptre in the world'.[81]

It is tempting to ascribe the deployment of foliage and greenery as a popular assertion of a common right to the natural produce of Ireland. However, in the absence of adequate sources 'from below', the individual motives of attendees

76 *Freeman's Journal*, 7 June 1843.

77 G. Owens, 'Visualizing the Liberator: Self-Fashioning, Dramaturgy, and the Construction of Daniel O'Connell', *Éire-Ireland*, xxxiii–xxxiv (1998), pp. 103–29. See also F. O'Ferrall, 'Daniel O'Connell, the "Liberator", 1775–1847: Changing Images', in R. Gillespie and B. P. Kennedy, eds., *Ireland: Art into History* (Dublin, 1994), pp. 100–103.

78 B. Jenkins, *Irish Nationalism and the British State: From Repeal to Revolutionary Nationalism* (Montreal, 2006), chapter 2 (p. 58).

79 *The Times*, 14 Sept. 1843.

80 *Freeman's Journal*, 25 May 1843.

81 *Freeman's Journal*, 17 August 1843. For background on Thomas Steele, see C. F. Lowth, 'Thomas Steele, Patriot Extraordinary', *Dublin Historical Record*, xliii (2000), pp. 92–118.

at the meetings are lost to the archival record.[82] Nonetheless, we can certainly assert that the co-option of physical manifestation of the natural environment as a national symbol acted as a practical means of evading proscription of the Repeal Association as a body.[83] And there can be no doubt that the act of using foliage to *create* an O'Connellite environment both served to demonstrate adhesion to the aspiration of repeal and also marked out those who refused to support it. At the Mullingar meeting, for example, it was reported that 'The entire front of every house was hid with green branches and wreaths of flowers, and the half-dozen residences that formed exception to the general rule only served to heighten the effect, in showing, by their bare walls (as destitute of the enlivening hue of the national colour as the hearts of their occupiers were of national feeling)'.[84] Similarly, there can be little doubt that when a triumphal arch made from evergreens and laurel was erected 'within a short distance of the gate of Lord Roden's demesne' it was intended as a message for the estate's notoriously ultra-Protestant owner.[85]

From their inception, the 'monster' meetings were about dominating public space in both small town and rural environments by juxtaposing peaceful O'Connellite crowds under the tutelage of the priesthood with symbols of state authority. At Trim, the event took place in the town square opposite the courthouse, whereas at Rathkeale the platform for the meeting was ostentatiously erected in front of the police barracks.[86] Interactions with nature were key to the popular iconography of the large numbers who were reported to have travelled from rural environments for the occasion, in multifarious acts of political pedestrianism: at Castlebar it was reported that 'The farmers and middlemen, of whom there were thousands in town, performed the journey on horseback, and in many cases carried their wives, sisters, or sweethearts behind them on pillions. The peasantry travelled on foot; but the distance and ruggedness of the roads had no disheartening effect on their spirits'.[87] Similarly, as the procession neared the racecourse near Ennis, County Clare, one reporter recorded 'interminable lines of human beings were seen across the country, rushing onward in eager haste to witness the glorious sight, and take a part in the imposing demonstration. Innumerable tents and marquees dotted the fields, as far as the eye could see'.[88] This processional aspect of the meetings entailed physical journeys between rural and urban environments

82 Cronin, '"Of One Mind?": O'Connellite Crowds in the 1830s and 1840s', p. 139.
83 N. Maddox, '"A Melancholy Record": The Story of the Nineteenth-Century Irish Party Processions Act', *Irish Jurist*, new ser., xxxix (2004), pp. 251–2.
84 *Freeman's Journal*, 16 May 1843.
85 *Freeman's Journal*, 1 July 1843.
86 *Freeman's Journal*, 20 April 1843.
87 *Freeman's Journal*, 2 August 1843.
88 *The Drogheda Argus and Leinster Journal*, 24 June 1843.

that acted as demonstration of numerical superiority: a typical *Freeman's Journal* account noted that:

> The entire population of the surrounding parishes, for a distance of some ten or twelve miles on every side, were to be seen crowding the streets of Trim from an early hour ... At short intervals along the entire road from the borders of the county of Dublin, triumphal arches and wreathes of laurel were erected, while even the poorest cabins had their door-posts ornamented with boughs of evergreen and garlands.[89]

For the landless, this was an interaction not just with nature and the environment, but also with property; as Angela Smith has noted, in many areas, 'Irish woodlands had long since disappeared and peasants burned turf in lieu of wood for fuel. Where trees were scarce, they were a valued commodity and an indicator of wealth and status. Only the gentry could afford to have woods or tree-lined avenues'.[90] It is thus instructive that leading priests took steps to 'impress upon their flocks the imperative necessity of obedience to the laws and deference to the authorities', and to ensure 'that the slightest injury should not be done to the plantations of the gentry—that not a bough should be cut without leave of the proprietor'.[91] What is equally striking is that these efforts failed to stem the use of foliage as a symbolic prop: the next week, at a meeting at Ennis, 'Men, women and children equally participated in the work of patriotism, and were to be seen entering the town from every direction, bearing on their shoulders trees and ever-greens procured from the surrounding country ... giving the town a most sylvan appearance'.[92]

Secular aspects of the 'monster' meeting should not be overstated. A majority of the meetings were held in dioceses in which the Catholic bishops had declared in favour of Repeal, and the Catholic clergy were deeply involved in the planning and execution of the 1843 campaign.[93] Their presence was both pervasive and highly visible, as the inclusion of forty-five priests among

89 *Freeman's Journal*, 21 March 1843.
90 A. Smith, 'Landscape Representation: Place and Identity in Nineteenth-Century Ordnance Survey Maps of Ireland', in P. J. Stewart and A. Strathern, eds., *Landscape, Memory and History* (London, 2003), p. 76.
91 The priest in question was the Revd Dr Collins, parish priest of Mallow, Co. Cork, assisted by 'the clergy of the town and neighbouring parishes', *Freeman's Journal*, 14 June 1843.
92 *Freeman's Journal*, 17 June 1843.
93 On the politicization of the Catholic Hierarchy in the period, see O. MacDonagh, 'The Politicisation of the Irish Catholic Bishops, 1800–1850', *Historical Journal*, xviii (1975), pp. 37–53; and on the various bishops' approaches to repeal, see D. A. Kerr, *Peel, Priests and Politics: Sir Robert Peel's Administration and the Roman Catholic Church in Ireland, 1841–1846* (Oxford, 1982), pp. 81–6.

the 358 persons admitted to the platform at Mullingar would indicate.[94] Dublin Castle received reports that priests were urging their flocks to attend, and some were going as far as to pay for carts to convey their parishioners to the place of meeting.[95] Most of the meetings were held on Sundays, and O'Connell publicly celebrated mass before the procession, on occasion joining it at the chapel gates.[96] So, too, did those who attended. O'Connell's son John wrote of the largest meeting, Tara, which was held on the Catholic Feast of the Assumption of the Blessed Virgin Mary, that 'Men came 40 & 50 miles to it; & slept on the hill all night; & had mass (it being the Feast of the Assumption) on the hill that morning'.[97]

There were signs, however, that other currents were at work. An anxious letter to Dublin Castle from the Protestant rector at Athboy recounted that he had seen a poster advertising the Tara meeting promising that the day's proceedings would begin with 'a solemn High Mass for the souls of the martyrs who were brutally massacred there in defence of their country's liberty in '98'. Exasperated, he asked simply: 'Is there a Government in Ireland?'[98] Furthermore, Dublin Castle was informed by a police constable that 'inhabitants of Carlow, Kildare, Wexford and Wicklow are to attend … some of these persons have a wish to see the graves of their fore-fathers who fell in battle at Tara in the year '98'.[99] *The Nation* reported that the Drogheda Trades, after their 'great national pilgrimage' to Tara, 'immediately proceeded to the centre of Rath-Na-Riodh, and planted their colours in a circle around the *Lia Fail*, which stands over the graves of the United Irishmen who fell at Tara in 1798'.[100]

The individual motives of those who engaged in such transgressive actions against O'Connellite constitutionalism are lost to history. We have no first-hand testimony of those who attended 'monster' meetings as rank-and-file supporters, or waved foliage in support of Repeal. Nonetheless, the campaign marked an extraordinary surge in constitutional pressure group politics in the pre-Famine period. Although attendance figures are impossible to calculate with any degree of accuracy, some insight is provided by close examination of the private information received by Dublin Castle. George Despard, who, as a military man may well have had experience of seeing crowds in anything like the numbers suggested, admitted the day after the Tara meeting that although it was 'wholly out of my power to form anything like an accurate

94 *Freeman's Journal*, 14 May 1843.
95 Report of Constable Courtenay, 27 July 1843, NAI CSORP 22/15/995.
96 See, for example, the meeting at Tullamore on 16 July. *The Nation*, 22 July 1843.
97 Journal of John O'Connell, 29 Aug. 1843.
98 R. Noble to E. Lucas, 5 Aug. 1843, NAI CSORP 22/15/763.
99 Constable J. Kinealy to E. Lucas, 1 Aug. 1843, NAI CSORP 22/15/995.
100 *The Nation*, 19 Aug. 1843.

calculation of the numbers present', nonetheless reported that 'I do not think I overrate it, when I say there were nearly 100,000 persons' present.[101] Owens has suggested that, even accounting for such wanton exaggeration, a quarter of the population of Leinster, Munster, and Connaught had taken some part in one of O'Connell's congregations.[102] This is probably an overestimation, but it should be noted that for the Government of the day appearances were just as important as reality: in May 1843, the Lord Lieutenant told the Prime Minister that it was the public perception of the numbers attending, as much as the numbers themselves, that was to be feared: 'Their numbers are we have reason to know are grossly exaggerated ... [but] ... when O'Connell states in public that he has just been addressing 70,000, 80,000, 120,000 men, the world believes him, & the effect upon the bold or the timid mind is proportionate'.[103]

The campaign itself was suspended when the Government prohibited a planned meeting at Clontarf and initiated prosecutions against O'Connell and other leading members of the Repeal Association. However, it is revealing that the symbolism of holding a demonstration at sites of Irish sovereignty was not lost on Peel's administration, and that it informed London's decision to act: the Chief Secretary told Downing Street that 'The meeting at Tara was very large & so far formidable more so *perhaps from the spot & the associated recollections* than from any peculiarity in the meeting itself'.[104] The O'Connellite 'monster' meeting involved the creation of a specific political environment and a popular interrelationship with a contested natural environment that can only be glimpsed in surviving sources. Nonetheless, the campaign's concatenation between Ireland's landscape, natural environment, and the case for self-rule would provide a rich furrow for the Young Irelanders—and others alert to the historical associations of place—to plough in succeeding years.[105]

101 G. Despard to E. Lucas, 16 Aug. 1843.
102 Owens, 'Nationalism without Words', p. 244.
103 De Grey to Peel, 6 May 1843. Peel papers, BL Add. MS 40478, fos. 39–44.
104 De Grey to Peel, 21 August 1843. Peel papers, BL Add. MS 40478, fos. 154–159.
105 T. Davis, 'Irish Scenery', first published in *The Nation*, 20 July 1844, in C. G. Duffy, ed., *Literary and Historical Essays by Thomas Davis* (Dublin, 1846), p. 211. See also J. Quinn, *Young Ireland and the Writing of Irish History* (Dublin, 2015), pp. 81–3.

# 5

# Therapeutic Environments in Nineteenth-Century Ireland: Hybrid Spaces and Practices

*Ronan Foley*

## Hybridity and Hydrotherapy

There were strong traditions in nineteenth-century Ireland that associated natural environments with aspects of health and healing.[1] Global interest in the interactions of the environment with human health has traditionally focused on disease, especially infectious disease,[2] as well as pollution and other elements of environmental justice.[3] Within historical research carried out by medical/health geographers, there has been a surge of interest in healthy places, broadly described as 'therapeutic landscapes'.[4] Therapeutic landscapes have been defined as 'a geographic metaphor for aiding in the understanding of how the healing process works itself out in places

1 See R. Foley, *Healing Waters: Therapeutic Landscapes in Historic and Contemporary Ireland* (Farnham, 2010) for detailed discussions of a range of settings going back to the pre-modern period.

2 Human–environment interactions that shape health can be seen explicitly in the writings of Hippocrates in the fifth century BC; see R. Porter, *The Greatest Benefit to Mankind: A Medical History of Humanity from Antiquity to the Present* (London, 1999), pp. 55–61.

3 A. Gatrell and S. Elliott, *Geographies of Health: An Introduction* (3rd edn., London, 2015); J. Adelman and F. Ludlow, 'The Past, Present and Future of Environmental History in Ireland', *Proceedings of the Royal Irish Academy: Archaeology, Culture, History, Literature*, cxiv (2014), pp. 359–91.

4 W. Gesler, 'Therapeutic Landscapes: Medical Issues in the Light of the New Cultural Geography', *Social Science and Medicine*, xxxiv (1992), pp. 735–46; R. Kearns and W. Gesler, eds., *Putting Health into Place: Landscape, Identity and Well-Being* (New York, 1998); A. Williams, ed., *Therapeutic Landscapes: The Dynamic between Place and Wellness* (Lanham, MD, 1999; Aldershot, 2007).

(or in situations, locales, settings, milieus)'.[5] In Ireland and elsewhere, they have been studied as a range of different natural/built environments where particular healing practices and narratives shaped the cultural production of place.[6] Equally, such spaces typically drew their healing reputations from specific environmental elements to be found locally. One especially important ingredient—water—and two specific forms of water-cure—spas and sweat-houses—are the central focus of this chapter.[7]

In considering such spaces and practices as hybrid, wherein blurred relations between nature and culture co-produce healing, we can frame them in relation to a number of multidisciplinary approaches including hybrid geographies,[8] contemporary writing on nature-based health,[9] and 'new topographers', whose work reflects a deeper mapping of well-being, place, and its human/non-human elements.[10] Within writing on environmental history, Sutter and others reflect this new interest in hybridity as a means, in part, to 'trouble' nature as a singular category, preferring a hybrid direction 'for approaches that see all environments as interweaving the natural and the cultural in complex ways'.[11] Adelman and Ludlow's short introduction to Irish environmental history is equally built on the hybrid interplay of three core themes: how environments influence human societies, how humans change environments, and their associated attitudes to the environment.[12] Foster and Chesney's first attempt at an environmental history of Ireland has less theoretical sophistication but equally draws from an interweaving of nature and culture to begin to understand how science, biology, and literature were all evident in how the Irish landscape was 'written up'.[13]

Water is an especially appropriate focus for historical studies of hybrid therapeutic environments. It is itself materially hybrid, but equally is a key

5 Gesler, 'Therapeutic Landscapes', p. 743.
6 J. Kelly, '"Drinking the Waters": Balneotherapeutic Medicine in Ireland, 1660–1850', *Studia Hibernica*, xxxv (2009), pp. 99–145; Foley, *Healing Waters*.
7 V. Strang, *The Meaning of Water* (Oxford, 2004); Foley, *Healing Waters*. Note this was also a reflection of the many global societies in which pharmacopoeias were developed from natural products. See Porter, *The Greatest Benefit to Mankind*.
8 S. Whatmore, *Hybrid Geographies* (London, 2002).
9 T. Hartig, R. Mitchell, S. de Vries, and H. Frumkin, 'Nature and Health', *Annual Review of Public Health*, xxxv (2014), pp. 207–28.
10 T. Robinson, *Stones of Aran: Pilgrimage* (London, 2008); H. Lorimer, 'Cultural Geography: The Busyness of Being "More-than-Representational"', *Progress in Human Geography*, xxix (2005), pp. 83–94; R. Macfarlane, *The Old Ways: A Journey on Foot* (London, 2012).
11 P. S. Sutter, 'The World with Us: The State of American Environmental History', *Journal of American History*, c (2013), pp. 94–119; J. W. Foster and H. Chesney, eds., *Nature in Ireland: A Scientific and Cultural History* (Dublin, 1997).
12 Adelman and Ludlow, 'The Past, Present and Future of Environmental History in Ireland', p. 361.
13 Foster and Chesney, *Nature in Ireland*; see Introduction.

shaper and element of natural landscapes.[14] At another level, it is central to healthy life, as the average human can survive much longer without food than water. The different forms and uses of water are also central to how it has been absorbed into practices of human health and healing—specifically, as discussed here, in hydrotherapy. There are multiple global examples of how water has been appropriated into human health practices via hot springs, holy wells, and early spa baths as well as through specific cultures of 'steam-cleaning', identified in settings like *hammams*, *saunas*, *banias*, and *onsen*.[15] The spa was a particularly resilient form, from its ongoing development in mainland Europe in the early thirteenth century through to its colonial dissemination in the New Worlds in the eighteenth and nineteenth centuries.[16] In almost all locations, including, as we shall see below, Ireland, a formalized eighteenth- and nineteenth-century culture emerged that produced very specific types of therapeutic landscapes for a predominantly wealthy clientele.[17] That generalized water-cure was reimagined and to an extent reinvented in the early mid-nineteenth century wherein an outburst of enthusiasm for new cold-water hydrotherapies led to the development of a range of specifically named hydros (hydropathic institutions) across Europe and the British Isles,[18] as well as a peculiar Irish revival of the Turkish Bath.[19] Yet more vernacular forms of the sweating cure existed in many countries and legacies of these too were found in Ireland.

In line with such a hybrid approach to nature and society, two contrasting nineteenth-century examples—the spa town and the sweat-house—will be used to demonstrate how specifically healthy environments were described, developed, and managed in Ireland. While the role of nature was crucial, it was shaped in turn by different social-cultural productions of place as well as their economic development. Both Irish examples reflected global healing narratives, yet identified that there were also local and epigenetic elements at play, specifically drawn from local cultures, places, and spaces. For geographers, there are important distinctions between place (as lived and

14 Strang, *The Meaning of Water*, pp. 3–4.

15 All sites of bathing cultures in the Islamic world, Scandinavia, Russia, and Japan respectively. See M. Aaland, *Sweat: Illustrated History and Description of the Finnish Sauna, Russian Bania, Islamic Hammam, Japanese Mushi-Buro, Mexican Temescal and American Indian and Eskimo Sweat Lodge* (Santa Barbara, CA, 1978); R, Porter, 'The Medical History of Waters and Spas: Introduction', in R. Porter, ed., *Medical History*, Wellcome Institute for the History of Medicine, Supplement No. 10 (London, 1990), pp. vii–xii.

16 The earliest spa was at Spa in Belgium, while other famous global examples included Baden-Baden, Vichy, Bagni di Lucca, Bath, Saratoga Springs, and Te Aroha. Porter, *Medical History*, Supplement.

17 Kelly, 'Drinking the Waters', 137–42; Foley, *Healing Waters*, pp. 50–1.

18 R. Foley, 'The Roman–Irish Bath: Medical/Health History as Therapeutic Assemblage', *Social Science and Medicine*, cvi (2014), pp. 10–19.

19 M. Shifrin, *Victorian Turkish Baths* (London, 2016).

emotional setting) and space (as locational and networked setting), but they can be read here as a combined descriptor for a functional geography. The choice of the two different study types is also important. While water was only part of the sweat-house cure, there were elements of bodily practices in nature that were a common theme across both settings. In addition to places/spaces, I would also argue that a site-specific practice/performative element has, to date, been underplayed in much environmental humanities work. In that sense, place and sociability were key components.

## Hybrid Natures, Cultures, and Health

Much of the scholarship in this chapter is informed by health geographies and it is important to consider where and how that research fits with theoretical discussions in environmental history; there are many commonalities, though equally some key departures. Sutter's review paper on US environmental history throws out many intriguing lines of development—some more wholeheartedly than others—on contested debates around hybridity, agency, causality, and complexity but also considers health in terms of human and natural environments.[20] He addresses a range of ideas that challenge the agency of nature, identify multiple competing narratives, and replaces environmental determinism with an uncovering of complex causalities. Of particular interest are some hybrid epidemiological terms he invokes, such as 'breathing space' and 'toxic bodies' that pick out specific harmful pollutants that have embodied effects.[21] Closer to home, Foster and Chesney's edited book is a first environmental humanities text for Ireland that, while driven primarily by a focus on flora and fauna, still concerns itself all through with nature–culture relations.[22] There are intriguing passing references in some of Foster's chapters to both *biophilia* and a hint that some attention might be given, in future research, to well-being in nature.[23] He also mentions herbs and folk medicine as documented in folklore texts,[24] but across this comprehensive text there is, overall, a limited interest in health–environment interactions. Adelman and Ludlow's recent paper on environmental history also shows an interest in aspects of environmental health and epidemiologies of disease.[25] They do identify health with specific reference to infectious disease and

20 Sutter, 'The World with Us', p. 96.
21 Ibid., p. 111.
22 Foster and Chesney, *Nature in Ireland*.
23 J. W. Foster, 'The Culture of Nature', ibid., p. 619 and p. 625 respectively.
24 J. W. Foster, 'Encountering Traditions', ibid., pp. 42–3.
25 Adelman and Ludlow, 'The Past, Present and Future of Environmental History in Ireland', pp. 382–3.

point towards oral history archives as a developing source for medical humanities, though beyond that the topic is mainly identified as a likely area for future work.[26] None of this should be especially surprising as medical/health history remains a relatively minority interest compared with other branches of the subject. Nonetheless, across work from environmental history one can glimpse a developing interest in narratives that consider healthy/unhealthy bodies and their relationship with a natural environment that has similar pathogenic/salutogenic components. To date, the former remains more prominent, but this chapter provides a more balanced discussion and considers the health-enhancing elements more fully.

Geographers too have a strong recent tradition in the study of hybridity. Such debates echo strongly in cultural geography writing on nature–culture, assemblage, actor–network theory, and non-human geographies.[27] Sarah Whatmore's key text, *Hybrid Geographies*, is especially relevant here. Drawing from Latour and others, Whatmore identified four areas of interest from which her research in hybridity emerges, namely theoretical work on non-representational theories, ANT (actor–network theory), corporeality, and what she calls 'the knowledge practices of everyday life'.[28] While none of these themes at first seem like they map onto environmental humanities, there is in all of them an implicit interest in complex place-based practices and how these are shaped by nature–culture.[29] Matthew Gandy's work is a good example of this writing, having a particular focus on water and its entangled relations within urban spaces, within which leisure and environmental health are important components.[30] Additional writing from the 'new topographers', noted above, provide complex narratives of different landscapes, from Essex and Suffolk to Connemara and Dartmoor,[31] that emphasis a mix of cultures, practices, and natures that have become equally attractive to cultural geographers.[32] Similar assemblages are identifiable in studies of therapeutic spaces in nineteenth-century Ireland like holy wells, but also in relation to the discovery of moorlands and coasts as outdoor healing spaces in the nineteenth-century

26 Ibid.

27 For a full list of around 1,000 detailed topics, see R. Kitchin and N. Thrift, eds., *International Encyclopedia of Human Geography* (12 vols., Oxford, 2009).

28 Whatmore, *Hybrid Geographies*, p. 6.

29 C. Brace and H. Geoghegan, 'Human Geographies of Climate Change: Landscape, Temporality, and Lay Knowledges', *Progress in Human Geography*, xxxv (2010), pp. 284–302; M. Gandy, *The Fabric of Space Water, Modernity, and the Urban Imagination* (Boston, 2014).

30 Ibid.

31 Robinson, *Stones of Aran*; Macfarlane, *The Old Ways*; R. Deakin, *Waterlog* (London, 2000); M. Kelly, *Quartz and Feldspar. Dartmoor: A British Landscape in Modern Times* (London, 2016).

32 J. Wylie, *Landscape* (Abingdon, 2007).

tourism and leisure industries.[33] Such invocations of nature's wider power of fresh air, greenery, and water, often developed relationally as an antidote to unhealthy urban living, echoed older narratives in the development of other forms of therapeutic landscapes including spas.[34]

With medical/health geography, research on therapeutic landscapes has had a strong historical dimension, mirrored in a more contemporary concern for what might be termed, 'nature-based health' and 'green and blue space'.[35] While early work focused on classical sites (Epidaurus), spa towns (Bath), or sites of pilgrimage (Lourdes), there was also a focus on the use of wilderness as a 'retreat' space, as well as more micro-scale studies of gardens and city parks.[36] Within this focus on natural green spaces, there was also an evident role for water in a variety of important nineteenth-century 'blue space' settings: spas, hot springs, hydros, seasides, and even sweating places and sites with spiritual water associations.[37] In considering access to identifiable sites of water-based healing, the role of nature as the provider of the raw ingredient was an important component. Yet nature on its own never made a functioning space and there was a strong human economic undertone in the social production of such places. In seeing such places as the co-produced result of culture overlaid on natural resources, there were evident materialities emergent from the environmental product—for example, minerals in water, the flow and availability of the healing product, as well as the use of wood, stone, sod, and water in creating built environments to take advantage of the healing powers of nature.

From these broad literatures, environmental humanities, social and cultural geography, and geographies of health, a number of specific themes will be explicitly discussed in this chapter, initially focused on the social and cultural performances/practices of health, within which nature and water played central roles. Spas, for example, did literally spring up out of the ground, but this was always a carefully choreographed process that started with the discovery of healing minerals and then stretched through scientific testing, medical confirmation of effective health outcomes, and a rapidly developing commercialization to a fully functioning local economy of healing.[38] Sweat

33 R. Foley, 'Performing Health in Place: The Holy Well as a Therapeutic Assemblage', *Health & Place*, xvii (2011), pp. 470–9; Kelly, *Quartz and Feldspar*, pp. 8–10.

34 Foley, *Healing Waters*, pp. 49–78; Kelly, 'Drinking the Waters', p. 137.

35 W. Gesler, *Healing Places* (Lanham, MD: 2003); Williams, *Therapeutic Landscapes*; Hartig, Mitchell, de Vries, and Frumkin, 'Nature and Health'; R. Foley and T. Kistemann, 'Blue Space Geographies: Enabling Health in Place', Introduction to Special Issue on healthy blue space, *Health & Place*, xxxv (2015), pp. 157–65.

36 Gesler, *Healing Places*; E. Palka, 'Accessible Wilderness as a Therapeutic Landscape: Experiencing the Nature of Denali National Park', in Williams, *Therapeutic Landscapes*, pp. 29–51.

37 Foley, *Healing Waters*; Foley, 'The Roman–Irish Bath'.

38 Foley, *Healing Waters*, Conclusion.

houses had different and much less studied histories, but essentially acted as epigenetic local spas, used and developed as communal healing sites with a significant embodied nature element. The remainder of this chapter will introduce spa towns and sweat-houses and then consider two themes to compare and contrast their functions as hybrid therapeutic environments in nineteenth-century Ireland. First, as landscape-based settings, they will be explored to identify the specific ways in which nature contributed to their material *identities* as therapeutic places. Second, the extent to which more *performative practices* cemented their social production as healing sites will be discussed. While the healing powers of the environment are more commonly considered in scientific/illness terms, an approach that considers the specific relevance of experiential and behavioural components places the idea of 'practice' more centrally in debates on the place-based production/promotion of health and wellness.

## Spa Towns and Sweat-houses

Spa towns are a global trope and date back to the twelfth century, with the earliest recorded 'business' associated with mineral waters recorded at Spa in Belgium. The Irish spa town, especially associated with a range of different mineral water types, had identifiable scientific characteristics for disease treatments, and were spread across the country from the earliest examples in the 1660s to late nineteenth-century revivals. Rutty's comprehensive list of Irish spa towns provided a clear classification of the seven different types of spa to be found in 1757, primarily sulphur and chalybeate (iron), while others included magnesium, saline, alkali, and vitriolic.[39] He identified around 120 named locations (including some holy wells), of which a small number were identifiable functioning spa towns.[40] Also listed were specific cure narratives, the temperature, taste, and behaviour of the water particles and its ability to 'bear carriage', as well as over eighty identified medical conditions for which spa waters provided treatments and cures (Figure 5.1). These ranged from cancer to consumption and from lameness to leprosy.[41] While spas generally declined in the face of a preference for seaside health resorts, key nineteenth-century examples included Lucan (rebuilt and redesigned at the end of the century), Ballynahinch, and Lisdoonvarna, which grew from almost nothing

39 J. Rutty, *An Essay towards a Natural, Experimental and Medicinal History of the Mineral Waters of Ireland* (Dublin, 1757); Kelly, 'Drinking the Waters'.

40 Key eighteenth-century spas included Wexford, Ballyspellan (Johnstown), Mallow, Templeogue, Swanlinbar, Lucan, Castleconnell, and Tralee.

41 Rutty, *Mineral Waters of Ireland*, Index Section I, pp. 474–6.

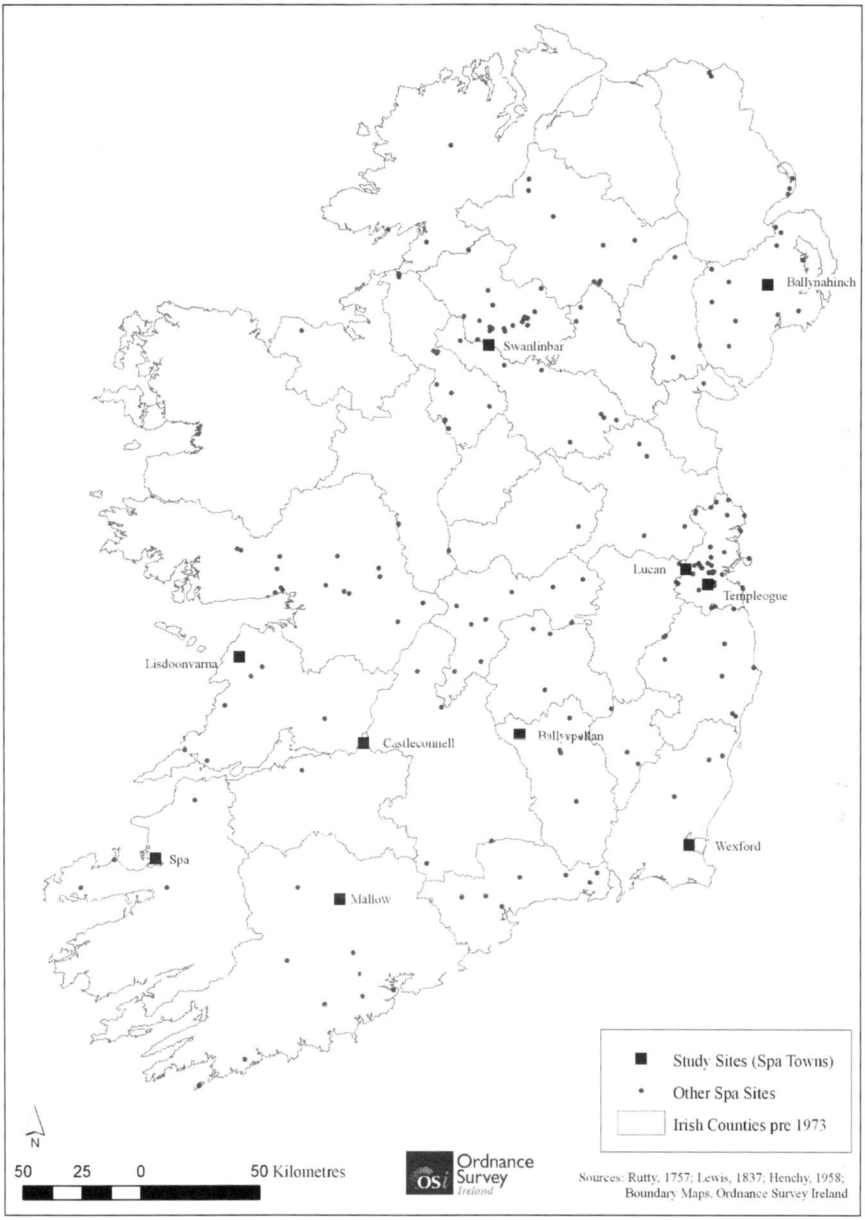


5.1 Map of Irish Spa Towns
Source: R. Foley, *Healing Waters: Therapeutic Landscapes in Historic and Contemporary Ireland* (Farnham, 2010)

in the 1840s to becoming the longest-surviving and most fully used by the end of the century.[42] Even at that point the sheer variety of spas and developmental potential were expressly noted in Lewis's *Topographical Directory*:

> On the demesne of Rathbane are several very powerful mineral springs, hitherto erroneously called the Lisdoonvarna spas, one of which is celebrated as being one of the strongest chalybeates in the kingdom ... found peculiarly efficacious in hepatitis, consumption, scorbutic and bilious affections, and rheumatism. Near this is another spring, which on analysis was found to contain, in addition to the iron, considerable portions of sulphur and magnesia ; the water is used with great benefit as an aperient. On the opposite side of a deep ravine, is a spring powerfully impregnated with naphtha, the exhalations of which taint the surrounding air; silver thrown into the water is instantly changed to a deep gold colour; and the water has been used with success as a cure for cutaneous diseases and for rheumatism. About a furlong further up the ravine, is a fourth spring, called the Copperas well; it has not been analysed, but has been used externally from time immemorial with effect as a cure for ulcers. The Rathbane mineral springs, under the appellation of the Lisdoonvarna spas, have been known and appreciated for centuries ... but from the bad state of the roads, and the want of proper accommodation, they have been comparatively neglected by invalids. Several cottages have, however, been recently built in the vicinity of these waters for the reception of visitors; and if the proprietor continues his improvements, and a facility of access be afforded, this place will probably become one of the most frequented spas in Ireland.[43]

There was a significant shift from the spa in the eighteenth century to that in the nineteenth century, with the latter more likely to have a spa doctor in place and a wider range of identifiable hydrotherapeutic treatments. Yet across their Irish history their efficacy was always contested by the rapidly developing professionalization of medicine and scientific specialization. Evidence of this contestation can be found in learned journals and also in literature–with Swift in particular having a history of ironic dissent in the eighteenth century, while Carleton provided one of the few fictionalized nineteenth-century accounts of

42 Kelly, 'Drinking the Waters'; R. Foley, A. Wheeler, and R. Kearns, 'Selling the Colonial Spa Town: The Contested Therapeutic Landscapes of Lisdoonvarna and Te Aroha', *Irish Geography*, xliv (2011), pp. 151–72; Rutty, *Mineral Waters of Ireland*, Index Section II, pp. 476–8.

43 S. Lewis, *A Topographical Dictionary of Ireland* (London, 1837), p. 183.

life at an Irish spa (Ballyspellan) that focused on satirizing both curative and social practices.[44]

At the other end of the scale, rural sweat-houses were found scattered across different upland parts of the country but were, from all available evidence, primarily concentrated in parts of North Connacht and Ulster (Figure 5.2). In contrast to the spa, they were a much smaller-scale, rural, and epigenetic form of therapeutic space, built from natural materials and used primarily to cure a variety of colds and arthritic conditions. Known in Irish as *teach alluis*, derived either from the Irish word for sweat, *allas*, or for cure, *leigheas*, they are in many ways intriguing and mysterious forms of a global sweat-cure that remain relatively unknown and unexplained. A range of quite fanciful propositions have been advanced as to their origins that involves everything from being a Viking remnant to practices brought back from American Indian cultures of sweat-lodges to being the real function of *fulacht fia*.[45] However, it seems from carbon dating and other evidence that they emerged in the early modern period and essentially had died out by the end of the nineteenth century, with the eighteenth and nineteenth centuries their period of peak use. They were mostly built in upland areas and were typically constructed from stone and sod. In appearance, they looked like igloos, with a low narrow entrance and a hole in the roof for steam to escape. Essentially, they were prepared well in advance, with turf fires which heated up the stones internally; patients then went in and sweated out their impurities in the blocked-up space. The next section provides a fuller description, but they arguably provided a form of local medical support in remote locations within which access to any sort of medic was limited. It has been argued that their decline was accelerated by both the impact of the Famine (they were generally built on marginal land) and the later nineteenth-century development of rudimentary primary and secondary care systems in rural areas.[46]

In carrying out empirical studies in both sites, key primary sources included scientific literature, travellers' accounts, diaries, Ordnance Survey letters, folklore archives, and photographs. The 1937 Schools Collection, given the bulk of the content within that collection of oral histories dates from the

44 E. Magennis, '"A Land of Milk and Honey": The Physico-Historical Society, Improvement and the Surveys of Mid-Eighteenth Century Ireland', *Proceedings of the Royal Irish Academy*, cii (2002), pp. 199–217; S. O'Sullivan, 'Jonathan Swift and Wexford's Spa', *Journal of the Old Wexford Society*, vi (1977), pp. 63–8; W. Carleton, *The Evil Eye or The Black Spectre* (Dublin, 1860), esp. chapters 20 and 21.

45 A. Weir, 1979, 'Sweat-Houses and Simple Stone Structures in County Louth and Elsewhere in Ireland', *Journal of the County Louth Archaeological and Historical Society*, xix (1979), pp. 185–97.

46 R. Foley, 'Leitrim's Sweat-Houses: Making Health from Nature', in B. Scott, ed., *Leitrim History and Society* (forthcoming); Foley, *Healing Waters*, pp. 79–110.

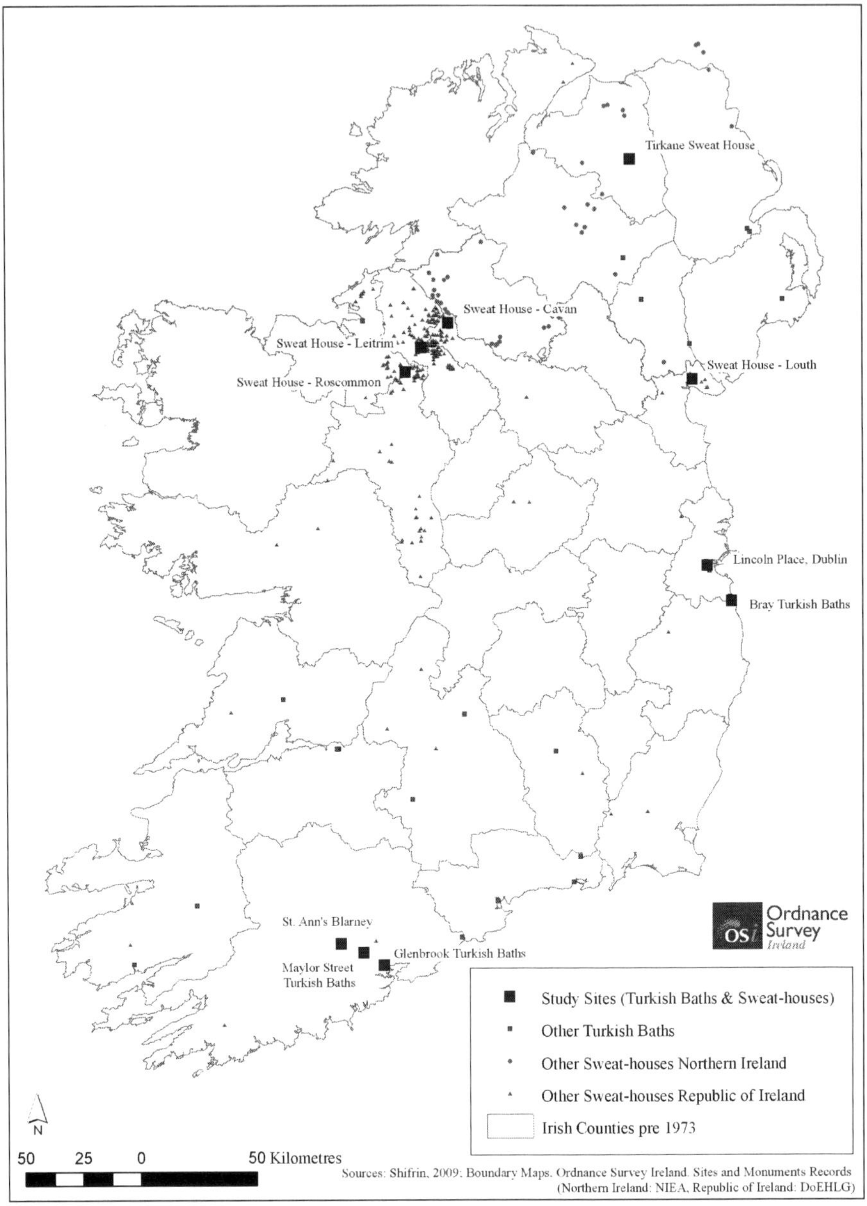


5.2 Map of Irish Sweat-houses
Source: R. Foley, *Healing Waters: Therapeutic Landscapes in Historic and Contemporary Ireland* (Farnham, 2010)

late nineteenth century, was particularly valuable for sweat-houses, and also for spa towns. Since its recent public dissemination online as an ongoing citizen transcription project (Duchas.ie), it is an especially rich resource for environmental humanities research because of its focus on themes including agriculture and rural life, customs, leisure, weather patterns, and, of most relevance here, healing.[47] The empirical evidence from spa towns and sweat-houses identified that they were both driven by combinations of heat, water, and bodies, and that each in turn embraced limited natural resources for their specific healing potential. But, equally, the primary material identifies the ways in which both settings were developed, inhabited, and perceived by a wide range of proponents and opponents, residents and visitors, and through the eyes of in-situ practitioners and passing strangers.

## Hybrid Natures: Landscape-Based Health and Wellness

In beginning to identify the specific ways in which nature contributed to their material identities as therapeutic landscapes, both spas and sweat-houses had at their heart a dependence on nature as provider. In the case of individual spas, one can see this best in the ways in which water drove such identities, such as at Lisdoonvarna, which in the mid-nineteenth century had one named well. By the end of the century, however, it advertised four different types of curative waters: copper, sulphur, chalybeate, and magnesia.[48] The discovery of what became the main sulphur well came after the others, the source allegedly discovered by a passer-by who picked up the faint 'rotten-egg' whiff of sulphur from a small stream, the Gowlaun, at the bottom of the town.[49] This sense of nature as source was central, and indeed Lisdoonvarna was relatively unique in the world in having two completely different water types, sulphur and chalybeate, within a few feet of one another at the Twin Wells (Figure 5.3).

Spa waters were shaped by different geologies, hydrologies, and often seasonal variations in the flow of the spring. As noted earlier in relation to Rutty's list, but also from contemporary research, there are intriguing links to an older variant, the holy well, with the same smells and often associated cures, identifiable at sites such as St Margaret's in North Dublin.[50] Equally,

47 For a full and detailed list of these topics as originally disseminated in 1937, see the Duchas.ie website: https://www.duchas.ie/download/17.01.26-irish-folklore-and-tradition.pdf.

48 Foley, Wheeler, and Kearns, 'Selling the Colonial Spa Town', pp. 7–8.

49 Foley, *Healing Waters*, chapter 3.

50 Rutty, *Mineral Waters of Ireland*, 299. Recent preliminary measurements by Bruce Misstear (personal communication) at Trinity College Dublin confirm elevated levels of magnesium at this site.

5.3 The Twin Wells at Lisdoonvarna.
The two different springs, sulphur and chalybeate, are both contained within the stone covering to the right, while the lady on the right, in charge of delivery, was a famous local character named 'Biddy the Sulphur'.
Image courtesy of the National Library of Ireland

specific hydrogeological aspects, such as the presence of warm springs, were identifiable natural components of spa water. The warm springs at Mallow were recorded as 68 °F by Rutty, and in later twentieth-century thermal surveys at around 19.5 °C (67 °F), though it was also noted that warm springs generally increased in temperature as discharge increased.[51] At Leixlip, warm springs were discovered during the building of the Royal Canal in the 1790s and a subsequent open air bath was built to take advantage of the warmish chalybeate waters, recorded at 75 °F (24 °C) at this site.[52] In truth, the Leixlip Spa was always inferior to the nearby spa at Lucan, a spa that had well-established seventeenth- and eighteenth-century origins, but which was extensively re-branded as a 'national' facility in the last decade of the nineteenth century.[53]

51 CSA Group, *Geothermal Energy Resource Map of Ireland: Final Report* (Dublin, 2004).
52 Foley, *Healing Waters*, p. 78.
53 Ibid.

The relationship between the chemicals in the water and their specific associations with a range of illness was at the heart of a nature–medicine relationship. Many of the beneficial effects were, ironically, founded on the difficulty that human digestive systems had in absorbing those chemicals, hence the 'purgative' nature of many of the cures.[54] Almost all narratives identified a wide range of illnesses treatable by each type of water. At Lisdoonvarna, it was noted that the Gowlaun Spring Water was 'most efficacious in cases of scaly skin diseases, in acne and eczema, in chronic rheumatism, rheumatic and atomic gout, scrofulous swellings, also in some diseases of bones and joints'.[55] The amount and frequency of the water one drank was also generally recommended by a combination of custom and the advice of spa doctors—of whom Rutty himself was one—to produce what might be termed a 'dose effect'. The same quote goes on to note, 'the proper dose of the Gowlaun water is from two to eight tumblers, or half pints daily, and the more divided this quantity the better, a stroll for some minutes intervening'.[56] However, in the later nineteenth-century spas one saw a slight move away from the traditional drinking and vomiting treatment regimens. While the same waters were drunk, there was a more scientific, even holistic, nature to, for example, the late nineteenth-century regimen at the recently designed spa at Lucan. Here the availability of fresh food and local produce, and indeed the value of nature for wider therapeutic exercises, such as walking, hunting, and golf, were considered important aspects of the cure.[57] At Lisdoonvarna, a series of new covered treatment rooms were developed at the beginning of the twentieth century, supervised by the spa doctor and delivered by what we might now call paramedics and complementary medical practitioners. The selling of these services was much more scientific and biomedical than natural, with advertising noting:

> The curative treatments now available and yearly being improved and supplemented, comprise, in addition to the hot and cold sulphur water baths, the needle bath, the general electric; the schnee bath (galvanic, faradic or combined), ionization and electro massage. Various medicated baths as prescribed, are also given, such as acid alkaline, bran, pine, peat, nauheim.[58]

In all of the Irish spas, the chemicals used and the illnesses treated were always identifiable as a local variant of a global form, reflecting the mid-century shift

54 Ibid.
55 P. Dwyer, *A Handbook to Lisdoonvarna and its Vicinity* ([1876] Ennis, 1998), p. 23.
56 Ibid.
57 T. M. Madden, *Lucan Spa and Hydropathic as a Modern Health Resort* (Dublin, 1891).
58 Foley, Wheeler, and Kearns, 'Selling the Colonial Spa Town', p. 9.

towards specific cold-water treatments and the later development of a wider set of embodied treatments including body wraps, showers, and massage.[59]

While sweat-houses were, on one level, much less developed, they still incorporated the use of natural constituents in their material healing practices. The ready availability of stone and sods especially, as well as the general presence of a stream or plunge pool to complete the treatment, were the fundamental natural building blocks of such healing settings. Their specific clustering in isolated upland areas was also loosely linked to limestone cover and specific soil types that made the extraction of blocky stone and grassy clumps, both central to the design, a product of specific local environments.[60] Wood-Martin's detailed description, below, lists their perceived derivation and form as well as how they were used operationally for curative purposes. Within this description one can identify a range of specific natural local materials that included stone, clay, turf, ash, rushes, and a plunge-pool (Figure 5.4).

> The hot-air bath, or what would now be designated the Turkish-bath, itself but a degenerate imitation of the luxurious laconicum of Ancient Greece and Imperial Rome—was in common use amongst the Irish, and in the County Sligo, lingered on until the commencement of this century. On the first map made of the county, several 'sweat-houses' were marked. They were generally of beehive shape, about 6 foot in diameter and 6 feet high, built of converging layers of monumental stones, covered with clay, and having a low narrow entrance, resembling the remains of huts still to be seen in juxtaposition with cashels or early ecclesiastical buildings, such as the hut styled the Sweat-house at Inishmurray. The nature of heating the chamber appears to have consisted of filling it with turf, igniting the fuel and when ... the ashes were almost out, and as soon as the floors and the side of the insides of the construction had sufficiently cooled down, the floor was covered in green rushes, the person or persons intending to take the bath entered the heated chamber, and the door was closed by means of a temporary screen. This hot-air bath was much used, not only for pleasure, but also as a cure for rheumatism, for which purpose it would seem to have been eminently successful. In some cases, it is stated that a pool of fresh water,—in the immediate vicinity—was utilised, as a plunge-bath for the perspiring bather, who remained in the heated ... as long as practicable, would then soak himself in the water and return.[61]

59 Ibid.; A. Durie, *Water is Best: The Hydros and Health Tourism in Scotland* (Edinburgh, 2006), pp. 88–9.

60 Foley, 'Leitrim's Sweat-Houses'.

61 C. Wood-Martin, *A History of Sligo: County and Town from the Close of the Revolution of 1688 to the Present Time* (Dublin, 1892), pp. 411–12.

5.4 Kiladiskert Sweat-house (exterior)
Source: Ronan Foley

He also suggests that the widespread use of sweat-houses had died off early in the century, though other oral history accounts dispute this, referring to their lingering use well up to the end of the century.[62] From specific local accounts on the Duchas.ie website, folk-memory recorded not just on whose land they were built, but also who built them:

> The first Sweat-house built in this locality was that in Corradeveret (now called Corrard) Parish of Killinagh.—The old lady who told me about this particular Sweat-house says it was built for her great-great-grandmother, a Mrs Carson. The man who built it was a Doctor Greden, he was a nephew of Mrs Carson's, and he recommended 'Sweating',—as a cure for Rheumatism, and all bone-aches.[63]

Sweat-houses were used for the treatment of a number of illness and conditions, many of which were associated with the rural environment. These included flu and respiratory illnesses, but also arthritis and rheumatism,

62 Tulach Cliabháin School: https://www.duchas.ie/en/cbes/4605943/4604490.
63 Cor na Gaoithe School: https://www.duchas.ie/en/cbes/4605944/4604565/4643467.

brought about by the rigours of agricultural labour outdoors.[64] In Leitrim, there are even recorded uses of bleeding at Murhaun, with the following account listing mid-nineteenth-century practices:

> The people suffering were also bled at the sweat house if sweating was not able to bring relief. The sweat house was beside the old house Terry Lynch was born in (John York's father). Turlogh as Terry was called could bleed and sweat but not as good as the Lynchs before him. The Lynches for generations bled people. If fever was supposed the sufferer was brought to the sweat house and if not relieved by sweating was bled.[65]

In this they mirrored wider sweating cultures worldwide such as saunas, *hammams*, and perhaps the nearest global comparator, *temazcalli* in Mexico.[66] Similarly, sweat-houses developed their curative power, literally, from the local environment linked directly to the ability of human bodies to improve health through sweating.[67] That indigenous healing knowledge drew from closeness to nature was evident in other therapeutic landscapes research from Canada, where First Nations groups described nature as the '24-hour pharmacy'.[68]

It is possible to see from both locations the raw ingredients of nature identified for a hybrid set of uses and cures with a range of different healing products. Some of these were pure, such as the drinking of a glass of sulphuric water, while others were more augmented forms, combined with other natural materials. The sweat-house was a co-production of stone, grass, water, and turf put to use in a relatively non-commodified way. In this, it was a distinct contrast to the spas, where that nature healing was made scientific, medical, and commercial, to produce what might be now seen as an early example of medical tourism. The power of nature was also narrated in very specific ways in such settings, with the spas in particular being highly dependent on how effective the stock of the 'twenty-four-hour pharmacy' might be to maintain the reputation of the therapeutic landscape. Yet such a role for narrative alongside nature hints at the importance of the social to this hybrid place production.

64 Foley, 'Leitrim's Sweat-Houses'.

65 Corderay School, Shancurry: https://www.duchas.ie/en/cbes/4605955/4605518/4643313?ChapterID=4605955.

66 A *temazcal* was a sweat-house that was made of the same stone materials and used for similar treatments, but was more horizontal in design and looked like a large pizza oven. They are still in use today.

67 Aaland, *Sweat*, pp. 132–3.

68 K. Wilson, 'Therapeutic Landscapes and First Nations Peoples: An Exploration of Culture, Health and Place', *Health & Place*, ix (2003), pp. 83–93.

## Hybrid Cultures: Socially Produced Health and Wellness

For both spa towns and sweat-houses, the natural environment was only one side of the story, and it is equally important to consider the extent to which more performative practices cemented their social production as healing sites. In essence, the natural product only provided one part of a curative culture that was, ultimately, socially produced. In Lisdoonvarna, for example, once the egg-like smell was identified at the Gowlaun spring, local landowners saw an opportunity, and called in, first, chemists, and then medics, to provide the scientific and medical evidence as to the volume and efficacy of the healing minerals in the water.[69] Once nature's healing powers had become 'authenticated' in this way, then a cycle of commodification kicked in, starting out with accommodation and moving on to a more performative set of practices which involved timetabled treatments and social mixing.[70] From the First Edition Ordnance Survey map in the early 1840s, one can see a reference to a single spa with a small cluster of houses. By the Second Edition, at the turn of the century, the town has developed significantly and now had six different named spas or bathhouses as well as a fully functional local therapeutic economy, built around the spa house, local assembly and treatment rooms, and *Maiville*, the associated spa doctor's house (Figure 5.5).[71] Across the town, new hotels, with evocative names like the Royal Spa and Hydro, had already been built to cater for the new markets, with the number of visitors recorded as 5,000 in 1878 and at least 20,000 in 1895.[72]

From the photograph, one can see glimpses of how the local landscape was incorporated into a space of healing performance. The nineteenth-century spa guides, at Lisdoonvarna and elsewhere, recommended drinking and walking in sequence and the more leisurely nature of spa society is evident in the image, with walkways and benches designed in alongside more natural slopes, trees, and streams. The building in the foreground was where the now pumped water was delivered (and still is in the summer season), while behind this were other buildings in which consultations and specific hydropathic treatment rooms were located. Performance at the spa extended beyond the typical daytime regimen of drinking and water, and in the evenings other social entertainments, dinner, and concerts, but also dancing and the distinctly unhealthy practice of drinking had a contested effect.[73] In the built environments around the spa waters, wider networking practices

69 Dwyer, *Handbook to Lisdoonvarna*.
70 Foley, *Healing Waters*, pp. 71–7.
71 Ibid.
72 Ibid.
73 Foley, Wheeler and Kearns, 'Selling the Colonial Spa Town', p. 15.

5.5 Lisdoonvarna Spa House (Maiville in background)
Image courtesy of the National Library of Ireland

produced a hybridity in the performances of health but also a hybridity in the ways in which patients responded to the regimens set out for them to produce perceived health benefits. In these wider social spaces of the spa town, culture and business were additional rituals that contested its reputational foundations as a solely natural site for health and healing.

Such commercial concerns and extensive social networking were less visible at sweat-houses, though they did have significant social roles and functions. Here, there was a less tamed version of nature, while the spaces were as much communal as social. There were no business opportunities and a limited role for medical professionals. There were accounts of local itinerant bath-masters who regulated who could or could not use the sweat-house; especially for patients who were overweight or had heart problems.[74] More commonly, sweat-houses were built by farmers and their neighbours, mostly for home consumption but also as a shared resource: anecdotally evidence suggested that in Leitrim and Sligo there was at least one sweat-house in every townland.[75] This extended to both shared materials and shared performances of care. Typically, men

74 Foley, 'Leitrim's Sweat-Houses'.
75 Ibid.

and women used the sweat-houses separately (with the exception of couples), while friends or family members acted as attendants to mind the clothes and provide blankets for when people needed to use plunge pools or streams after the sweating stage was over. Here, the performative elements were distinctly embodied ones, with a functional healing focused primarily on making oneself fit to work again. Yet there were also accounts of more cosmetics uses, with sweat-houses on Rathlin Island used by young women to improve complexions and to clean away the stains left by agricultural labour such as flax picking and seaweed gathering, especially when preparing to attend potential romantic settings such as fairs and markets.[76]

## Hybrid Therapeutic Geographies in Nineteenth-Century Ireland

> They [spas] demonstrate the degree to which past cultures of health were complex performances—enterprises shared between the sick and the medical profession (itself intricately stratified); operating within a matrix of resources, institutions, amenities and physical buildings; and drawing upon elaborate rituals of regime ... They lasted ... because they satisfied a deep desire that the healing enterprise should proceed within frameworks essentially sociable in their nature, and suffused with symbolic cultural meanings.[77]

The existence of historical cultures of health in nineteenth-century Ireland has to date been dominated by the experience of illness rather than the ways in which environment and society operated together to produce and maintain health. While there is no doubt that the Famine left a deep scar, the result was the development of rudimentary secondary and primary health care systems in the form of dispensaries, county hospitals, asylums, and rural doctors.[78] Yet older variants of therapeutic landscapes, in which nature provided the healing ingredients, were also visible across the century. In seeing these spaces as hybrid, two different strands emerge in the paper, one focused on the hybrid material environments, the other on the hybrid practices/performances that together produced a reputational healing in place.

More widely, it is also important to consider how these different forms of hydrotherapy in the later decades of the nineteenth century mirrored debates about the relationships between orthodox and unorthodox medical practices.

76 Aaland, *Sweat*, pp. 132–3.
77 Porter, 'The Medical History of Waters and Spas', p. xii.
78 C. Cox and M. Luddy, *Cultures of Care in Irish Medical History, 1750–1950* (Basingstoke, 2010).

One might see this as also marking off a subtle shift in curative and medical practices from nature to science. While that delineation was never initially clear, older, more-blurred practices that one might have associated with folk medicine were becoming more sharply delineated and governed. While it might be a stretch, one could see the decline of both spas and sweat-houses as representing a sort of 'de-naturing' of healing orthodoxies, which makes a revived modern interest in complementary and alternative medicines an ironic contemporary return to older healing practices.[79]

While the environmental scale was small, both the spa and sweat-house might be seen as variants of what Hoyez refers to as 'globalising therapeutic landscapes',[80] and were reflective of a wider global history of the social production of certain curative forms and cultures. While associations with nature tend to frame them as outdoor spaces, the locations of both spas and sweat-houses included a hybrid mix of outside–inside space. Within spas, there was a movement between the inside and outside as part of a designed treatment interaction: drinking, walking/promenading, healthy water, and air all combined together to provide healing benefits. Yet that inside/outside element was also central to the sweat-house, with a sort of enhanced nature within the hot interior contrasted in very material and embodied ways with the plunge pool outside. Indeed, wrapping up warm, before and after the treatment, was considered crucial in the completion of the cure.[81] Both settings might arguably be seen as assemblages of health and nature, within which multiple actants (human and non-human) engaged in complex relational acts that blurred the boundaries of nature and culture; indeed, one might use the term 'healthscapes' as a synonym for therapeutic landscapes, to emphasize that connection.

That nature–culture hybridity was enhanced in specifically performative ways, which was a feature of healing water histories in Ireland and seen in other curative settings such as holy wells and Turkish baths.[82] The two forms had variable histories across the century that reflected global/local narratives, yet differed in crucial ways around their recognition and perceived efficacy as healing spaces. The performance of health at both settings had strongly embodied components and used different hot/cold forms of hydropathy (as another example of the hybrid ways in which water was co-opted as a curative element) in very different settings, within which hybrid roles for nature and the environment were experienced and played out in different ways.

79 Foley, *Healing Waters*; see Conclusion.

80 A.-C. Hoyez, 'The "World of Yoga": The Production and Reproduction of Therapeutic Landscapes', *Social Science and Medicine*, lxv (2007), pp. 112–24.

81 Foley, 'Leitrim's Sweat-Houses'.

82 Foley, *Healing Waters*.

In comparing spas and sweat-houses, the former might be characterized as a culturally produced performances of health versus a more pragmatic/natural performance at the latter. Equally, one might contrast the spa and the sweat-house as, respectively, special and everyday places. Both were also liminal places and here one might see that liminality as part of a contested and hybrid identity. Healthy and unhealthy behaviours were enacted side by side at the spa: at Swanlinbar, for every morning glass of sulphuric water, a glass of claret was consumed in the evening.[83] Other health-endangering behaviours, such as not wrapping up properly, were also noted at the sweat-house.[84]

In exploring these two examples of therapeutic landscapes and especially their functions as places providing different forms of health care treatment, they reflect Sutter's suggestion that such spaces act as, 'an argument for cultural and environmental contingency rather than genetic determinism'.[85] The health outcome of the interaction between man and nature was never guaranteed, and indeed one might even see the results in hydrological terms as a space in which both clean/unclean water and bodies intermingled. This also echoes Kelly's contrasting of separate environmental histories of clean water in Dartmoor with the contrasting muddied water of the spa.[86] There are other contemporary connections as the material remnants of both spas and sweat-houses still sit in the Irish landscape, as forms of under-appreciated cultural heritage but equally as early examples of contemporary spaces within which that same nature/culture link survives. While sweating cultures continue to thrive worldwide, it is also interesting to note that contemporary spas, wellness spaces, and embodied healing treatments continue that focus on a designed nature/culture connection, usually in built environments that explicitly seek to 'bring the outside in'. In an attempt to historicize nature, links to newer research on healthy natures and blue space[87] can be usefully traced back to its nineteenth-century antecedents while a specifically therapeutic framing of nature/culture debates and hybrid natures has plenty of room for development in our understanding of present-day landscapes and ecologies.

83 Ibid., p. 63.
84 Foley, 'Leitrim's Sweat-Houses'.
85 Sutter, 'The World with Us', p. 113.
86 Kelly, *Quartz and Feldspar*, p. 276.
87 Foley and Kistemann, 'Blue Space Geographies'.

# 6

# On Why the UK's First National Park Might Have Been in Ireland

*Matthew Kelly*

The UK government first gave serious thought to designating National Parks in 1929 when Ramsay MacDonald's Labour Government commissioned Christopher Addison to investigate the issue. His recommendations were received favourably in Whitehall, but their formal adoption was derailed by the economic crisis engulfing the country. The successor National Government paid lip service to Addison's proposals, but significant executive action was not taken. The issue was revisited during the Second World War and taken up by the Attlee governments. The National Parks Act 1949 established the National Parks Commission, which was empowered to designate National Parks in England and Wales. By 1957, ten were designated. Much of the drive for these designations came from access campaigners and it was no coincidence that the legislative steps were taken by a Labour government. National Park designation was as integral to the social democratic agenda as the creation of the National Health Service.[1]

As is often the case in British history, a broadly accepted historical narrative is complicated if account is taken of Ireland. National Park history in the United Kingdom would look very different had lobbyists succeeded in persuading the government to purchase the Muckross Demesne in County Kerry and manage it as a National Park when it came on the market in 1899. The same might be said of the history of the National Trust, which launched a campaign to raise the funds to buy the estate. Such a large holding, dwarfing any other National Trust acquisition of the period, would have reorientated the geographical focus of the organization and given it a genuine British–Irish character. Either form of acquisition would also have positioned Ireland more

1 John Mair and John Delafons, 'The Policy Origins of Britain's National Parks: The Addison Committee 1929–31', *Planning Perspectives*, xvi (2001), pp. 293–309.

prominently within two larger historical processes, namely what one historian has called 'the dawn of green' and others have characterized as the history of the nature state.[2]

That the government refused to step in, and that the National Trust campaign made little headway, probably accounts for why this episode has been largely overlooked by historians.[3] It deserves revisiting for a number of reasons. Most obviously, the question had political traction because the estate included much of what was generally considered the most charismatic landscape in Ireland and, according to many commentators, in the UK. This not only gives the episode significance in the history of landscape appreciation and notions of natural beauty, but also in the history of the open spaces movement and the challenge it posed to private property. Lobbyists and campaigners, having learned that central government was reluctant to protect legally indefensible customary rights of access, had become enmeshed in a new politics of purchase as the only sure way to secure public access to charismatic landscapes. London County Council's decision to buy Epping Forest on behalf of the people of London in 1882 was a key moment in this history, while in the 1890s there were calls for Devon County Council for purchase the whole of Dartmoor in order to protect it for the people of the county from predatory London water undertakers.[4] Better known is the establishment of the National Trust in 1895, a response by a group of well-placed activists to the need for a body that could coordinate purchase campaigns, accept legacies, and manage property in the public interest. In general, preservationism and the politics of access reflected a radical and democratic agenda predicated on the notion that certain charismatic landscapes, in the famous phrase of Wordsworth, were 'a form of national property'.[5]

Ireland's relationship to this sense of the nation's collective patrimony was complicated. As this essay will show, the Muckross controversy allowed Home Rule MPs to highlight the inequities of the Union and the shortcomings of the British state in Ireland. Fresh, if short-lived, impetus was given to the 'financial relations' question that dogged Anglo-Irish relations since the Act of Union, especially since the advent of Gladstone's home rule bills. The Royal Commission appointed to examine the issue concluded in 1896 that

2 Harriet Ritvo, *The Dawn of Green: Manchester, Thirlmere, and Modern Environmentalism* (Chicago, 2009) and W. Graf von Hardenberg, M. Kelly, C. Leal, and E. Wakild, eds., *The Nature State: New Approaches to the History of Conservation* (London, 2017).

3 The exception is K. J. James, *Tourism, Land and Landscape in Ireland: The Commodification of Culture* (New York, 2014), to which this chapter is at times indebted.

4 M. Kelly, *Quartz and Feldspar. Dartmoor: A British Landscape in Modern Times* (London, 2016), pp. 211–18.

5 Paul Readman, *Storied Ground: Landscape and the Shaping of English National Identity* (Cambridge, 2018).

Ireland's tax revenue was about one-eleventh of the UK total but its 'relative taxable capacity' was about one-twentieth. That historians have found this a dubious calculation did not diminish its immediate political significance.[6] The state's potential acquisition of Muckross was presented by Home Rule MPs as both a modest act of reparation and as evidence of the government's sincerity in its professed determination to govern Ireland justly. The US example loomed large in their polemics. The federal government's decision to acquire and designate Yellowstone National Park in 1872 and its acquisition of further parks—Yosemite National Park followed in 1890—was presented as evidence of what a genuinely democratic state could achieve on behalf of its people. The question of who constituted the people gave the issue added traction. Killarney's boosters argued that the region's transcendent natural beauty meant it constituted a form of transnational commons, a claim that served the interests of the tourist industry and could be given a nationalist or a unionist inflection. Nationalists tended to claim that Killarney should be held in trust by the Irish people on behalf of humanity, whereas unionists, particularly those associated with the National Trust, considered Killarney to be threatened by American investors insensitive to its importance to the patrimony of the UK as a whole. In either case, the challenge both parties offered to conventional notions of private property and the remit of the state could only be sustained if Killarney's exceptional natural qualities were seen to make this landscape deserving of a new form of political status.

Finally, to scroll ahead into the twentieth century reveals how the resolution of the Killarney question saw first the Irish Free State and then the Republic of Ireland adopt the US rather the UK national park model. In contrast to the US model, which is predicated on state ownership, the British model extends enhanced planning powers to the national park authorities. It would be mistaken to suppose that this difference means the performance of the Irish nature state has been more impressive than its British counterpart. This is a relative claim, of course, for both systems have faced severe criticism, but it can hardly be doubted that a legacy of British rule in Ireland has been the pronounced reluctance of Irish legislators to empower the Irish nature state at the expense of the agricultural or landowning interest. To revisit the Killarney debate of the late nineteenth century from the perspective of the twenty-first is to be struck less by its explicit radicalism and more by the implicit claim

6 On the complexity of the issue, see L. Kennedy and D. S. Johnson, 'The Union of Ireland and Britain 1801–1921', in D. George Boyce and A. O'Day, eds., *The Making of Modern Irish History: Revisionism and the Revisionist Controversy* (London, 2006), pp. 44ff.; W. E. Vaughan, 'Ireland c.1870', in Vaughan, ed., *A New History of Ireland*, V: *Ireland under the Union, I: 1801–1870* (Oxford, 1989), pp. 784–94; H. D. Gribbon, 'Economic and Social History, 1850–1921', in Vaughan, ed., *A New History of Ireland*, VI: *Ireland under the Union, II: 1870–1921* (Oxford, 1989), pp. 330–1.

that the natural characteristics of Killarney's mountains and lakes made them an environment apart. The real business of the Irish countryside was not conducted on mountainous, stony ground.

*

A number of historians have traced the development of Killarney's exceptional status as a charismatic landscape from the middle of the eighteenth century, particularly through the study of Killarney tourism and romanticism.[7] Early proto-romantic responses to Killarney include Lord Orrery's musing in a letter of 1735 and an anonymous work of 1750 that presented Killarney as a 'Masterpiece of Nature'.[8] In the decades that followed, descriptive works proliferated in the form of poetry, public letters, or itineraries, often recounting travellers' experiences, plus the occasional novel and books of engravings and of 'lore', 'sayings', and 'legends'.[9] Arthur Young, visiting in September 1776, declared that 'among the Lakes that I have seen', Killarney 'can scarcely be said to have a rival'; Lady Chatterton and Sir Richard Colt Hoare were among the prominent commentators to follow suit. Lake, mountain, and woodland combined to provide sensitivities schooled by the dictums of romanticism and the picturesque visual cues that proved deeply satisfying. Killarney's boosters keenly quoted these authorities,[10] and by the 1820s the region was marketable as a tourist destination. The first 'guide' was published in 1822, though most influential was probably Mr & Mrs Hall's *A Week at Killarney* (London, 1843), which was published in a new edition in 1865.[11]

By the middle of the nineteenth century, a clear visitor itinerary had emerged, and Killarney was, and would remain, notorious for the degree to which local people sought to exploit the popularity of the district. In the pithy summary of Kevin James, 'Killarney's supposed wretched misfortune, as well as its signature draw, was the remarkable peasant pantomime for which it

7 In addition to James, *Tourism, Land and Landscape*, see, in particular, W. H. A. Williams, *Creating Irish Tourism: The First Century, 1750–1850* (London, 2011).

8 L. Gibbons, 'Topographies of Terror: Killarney and the Politics of the Sublime', *South Atlantic Quarterly*, xcv (1996), p. 25.

9 These are too numerous to list, though a search for 'Killarney' in the digital catalogue of either the National Library of Ireland or the British Library yields much.

10 See, for example, 'Description of the Lakes of Killarney and the Surrounding Scenery' (1849), pp. 48–53.

11 For example, the work of the Halls was deferred to by 'One of the Rakes of Mallow' in 'Killarney—How to See It, and Where to Fish in It', *New Sporting Magazine*, Aug. 1856, p. 142.

served as a backdrop'.[12] The scathing review *The Athenaeum* gave *Paddy-Land and the Lakes of Killarney* (1853), by 'a Wykemist', a pamphlet intended to promote touring in Ireland, was indicative of just how clichéd such treatments had become.[13] More respectable than the rituals demanded at Kate Kearney's cottage, where travellers were expected to take a drop of the poteen, was the thriving trade in Killarney woodware. Manufactured from bog oak or arbutus, offerings ranged from small mementoes for tourists to high-quality items of furniture whose distinctive marquetry depicted images of iconic Killarney sites like Muckross Lake, Glena Cottage, and Ross Cottage as well as ferns, mountain eagles, and deer.[14] Other literary artefacts testifying to Killarney's exceptional status include Michael Doheny's *The Felon's Track* (1849) and Dion Boucicault's *The Colleen Bawn* (1860). The arrival of the railway in 1853 further catalysed the development of Killarney's tourist infrastructure; the status of the region was secured by Queen Victoria's visit in 1861, during which she spent four nights at Muckross House, the fine 1830s centrepiece of Henry Herbert's Muckross Estate. The Queen 'ceremoniously traced the beaten track', but the debt-financed improvements implemented by Herbert, including a new road on the slopes of Mangerton, was not in the long-term interest of the estate—and failed to secure Herbert a knighthood.[15]

British enthusiasm for Killarney did not pall. Alfred Austin's long account of his tour of Ireland, run in *Blackwood's* in August 1894, included rapturous descriptions of Killarney—'there is nothing in England and Scotland as beautiful'[16]—and prestigious visitors continued to make the journey west. Hot on the trail of the Royal Scottish Aboricultural Society was the Duke and Duchess of York, who stayed with Lord Kenmare during their tour of Ireland in August 1897. Their visit to neighbouring Muckross and its famous sites was marred by rain but revived persistent rumours that the demesne would be bought as the royal residence in Ireland.[17]

The Herberts certainly needed bailing out. By the 1890s, renewed attempts to render the estate solvent, including reviving its mining

12 James, *Tourism, Land and Landscape in Ireland*, p. 10. See also G. Baylis, 'Exchanging Looks: Gap Girls and Colleens in Early Irish Tourist Photography', in *Early Popular Visual Culture*, x (Nov. 2012).

13 *The Athenaeum* (20 Aug. 1853).

14 B. Austen, 'Tourism and Industry: Killarney and its Furniture', *Irish Arts Review Yearbook*, xii (1996), pp. 45–55.

15 James, *Tourism, Land and Landscape in Ireland*, 41; on the Herberts and Muckross, see S. McCoole, 'The Herberts of Muckross', in J. Larner, ed., *Killarney: History and Heritage* (Wilton, 2005), pp. 90–110; D. S. Lucey, *Land, Popular Politics and Agrarian Violence in Ireland: The Case of County Kerry, 1872–86* (Dublin, 2011), pp. 19, 51–2, 155.

16 Alfred Austin, 'That Damnable Country', *Blackwood's Edinburgh Magazine*, clvi (Sept. 1894), p. 318.

17 *Freeman's Journal*, 10 June 1895, 26 Aug. 1897, 30 Aug. 1897.

industry—echoes here of Parnell's work at Avondale—had come to naught and by the 1890s the estate was heavily mortgaged. Standard Life initiated foreclosure proceedings. Fears public access might be prohibited were immediately voiced, as was the claim that the sale invited state intervention. 'In any other country in the world the State would interfere and purchase Muckross as a National Park', insisted the *Freeman's Journal*, arguing that the UK government should emulate its American counterpart, which 'with characteristic magnificence' had 'bought up vast territories containing thousands of square miles and dedicated them forever to the public'.[18] The Swiss government's treatment of certain Alpine landscapes offered another inspiring parallel, as did the state's maintenance of London's parks and London County Council's acquisition of Epping Forest. The 'state', it seemed, did not simply have the capacity to deliver what might now be termed a form of environmental justice, but, as transnational comparisons suggested, was so obliged.

A month or so later, the *Freeman's Journal* followed-up with a wide-ranging and highly polemical editorial. It struck an internationalist note, insisting Killarney, 'a shrine of Nature's loveliness', did not 'belong to Ireland alone'; it needed to be protected against the 'profit-making vandalism' of this 'utilitarian age' and the dangers posed by a 'syndicate' of investors. The threat was comprehensive. Killarney's timber would be felled, Torc Waterfall would be harnessed to 'a turbine wheel … to provide electric energy for the manufacture of carbide of calcium', and the House and possibly the Abbey would be converted into a chemical factory and the Lakes into a 'vacation resort, according to Cockney notions, a rival, say, of Margate'. The fate awaiting the islands was conversion into tea gardens serviced by a fleet of steam launches. Subject to the depredations of industrial modernity and its attendant leisure cultures, Killarney would cease to inspire painters and poets, its 'Nature' no longer a 'sermon … telling of the power of God and the wonder of this work of His hands'.[19]

The tastelessness of the upwardly mobile working and middle classes of London was an established trope in preservationist polemic, but to recognize this does not in itself make the *Freeman's* fears outlandish. Concern about the malign power of syndicates was characteristic of the period, and more specifically the Killarney debate was conditioned by the purchase of the lease to the Giant's Causeway by just such a 'syndicate' in 1896. Railings had been erected, tolls introduced and the Portrush hydroelectric tramway extended. Thomas McKnight disparaged these developments as 'cockneyfying', but the government had not considered the enclosure of the site a threat to the

18 *Freeman's Journal*, 7 Apr. 1899.
19 *Freeman's Journal*, 12 May 1899.

public interest.[20] Moreover, calcium carbide, invented in 1892 and used in steel production and as the agent in acetylene gas, was mass-produced using hydroelectric power at Niagara Falls, though it is likely the *Freeman* was also aware that in 1896 the North British Aluminium Company won the right, in the face of considerable opposition, to harness the Falls of Foyer at Loch Ness in Scotland for the same purpose.[21] The parliamentary Select Committee appointed to consider the Industries (Ireland) Bill 1897 had been informed that calcium carbide could be produced in Ireland where there was plentiful supply of lime and cheap land.[22]

Home Rule MPs asked questions in the House of Commons. Did Arthur Balfour, the First Lord of the Treasury, intend to 'take steps to secure this property for the people of Ireland'? Balfour said no. Might the estate be purchased with the '100 millions due to Ireland' or the 'still undiscovered remnant of the Irish Church Fund'? Again, no. Thomas Lough, Liberal MP for Islington West, but an Anglo-Irishman from Cavan, was more conciliatory, asking if there was a 'fund for the preservation of national and historic monuments' which might be drawn on. Balfour responded that the estate was of neither national nor historical significance.[23] A letter to *The Times* also made the case for state intervention. The author's warning that a new owner could cut off access to Killarney's most famous sites, including the lakes, and cut down the demesne's forests, opening them up to grazing, linked the Muckross question to issues animating agrarian and access politics. Further observations posited an extended role for the state in Ireland as a necessary response to the country's peculiar circumstances. It was not simply that this 'most valuable asset of the kingdom of Ireland', so important to the country's tourism, could not be secured as a public park, but that Killarney—and Ireland more generally—was peculiarly vulnerable to unscrupulous private speculators because its social and economic elite did not fulfil its paternalist obligations. The plight of Muckross was yet more evidence of the dysfunctional nature of the Union. Only the Guinness family, it was claimed, could be relied on, which meant the state must be ready to take up the slack; an Irish parliament, the correspondent claimed, would not hesitate to bring Killarney into public ownership.[24]

20 Kelly, *Quartz and Feldspar*, 214; Irene Furlong, 'The Landscape for All—No Penny-in-the-Slot at the Giant's Causeway', in Ú. Ní Bhroiméil and G. Hooper, eds., *Land and Landscape in Nineteenth-Century Ireland* (Dublin, 2008), pp. 71, 74–6.

21 A. Perchard, *Aluminiumville: Government, Global Business and the Scottish Highlands* (Lancaster, 2012); Janice Helland, 'Artistic Advocate: Mary Rose Hill Burton and the Falls of Foyers', *Scottish Economic and Social History*, xvii (Nov. 1997), pp. 127–47.

22 *Morning Post*, 10 April 1897.

23 Hansard, HC vol. 71, col. 748 (16 May 1899).

24 *The Times*, 26 May 1899.

Lough's proposal that Irish funds might be used to purchase Muckross demesne 'as a national park', was not the first time the term 'National Park' had been used in parliament.[25] Recurrent parliamentary debates about how London's Hyde Park, St James's Park, and Green Park were financed, which fell on the exchequer rather that London County Council, toyed with the idea that their special status made them 'national parks', but this was a rather different claim to that made of Yosemite, Yellowstone, or Niagara.[26] Similarly, during the debate on the Ambleside Railway Bill in 1887, James Bryce claimed that 'the people look upon the Lake District as their national park, and they desire to preserve it', but although he took inspiration from the treatment of Yosemite and Niagara Falls by the US and Canadian governments, he was confident that existing parliamentary powers, properly exercised, provided adequate protection.[27]

Balfour framed his response according to unionist principles, insisting that no distinction could be drawn between 'Imperial money' spent in Ireland or England. This directly contradicted the premise of the 1894 Royal Commission, which treated Britain and Ireland fiscally 'as separate entities', but was of course consistent with the broad assumptions underpinning the Union.[28] Balfour also thought there was no logic to buying a park 'so far from any great centre of population' which would have 'little value to the public during many of the winter months of the year'.[29] Kerry County Council and other public bodies joined the fray, claiming that if the government purchased the estate and managed it as a 'National Park', 'the income derivable from the property' would ensure 'the loss to the revenue would be inconsiderable, if any'.[30] Balfour again rejected the suggestion, adding that there was no evidence that public opinion throughout the UK favoured such use of public funds.[31] It was a case for the 'benevolence of private individuals'.[32] When Michael Hicks-Beach, chancellor of the exchequer, was challenged by nationalist MPs about negotiations between the Commissioners for Woods and Forests and the Duke of Beaufort for the purchase of the Raglan Estate and Tintern Abbey, he responded that, in contrast with Muckross, the Raglan Estate was a profitable agricultural enterprise.[33] James Winter suggests that the purchase

25 Hansard, HC vol. 71, col. 748 (16 May 1899).

26 Hansard, HC vol. 287, col. 235 (21 Apr. 1884); vol. 303, cols. 494, 496–497, 503 (11 Mar. 1886); vol. 334, cols. 1083–1084 (28 Mar. 1889).

27 Hansard, HC vol. 310, cols. 1734–1735 (17 Feb. 1887); see Readman, *Storied Ground*, p. 109ff.

28 Kennedy and Johnson, 'The Union of Ireland and Britain', p. 44.

29 Hansard, HC vol. 72, col. 1501 (19 June 1899).

30 *Freeman's Journal*, 30 May 1899 and 10 July 1896.

31 Hansard, HC vol. 73, cols. 75–76 (20 June 1899).

32 Hansard, HC vol. 73, cols. 456–457 (23 June 1899).

33 Hansard, HC vol. 73, col. 764 (27 June 1899). The comparison had been raised by the *Freeman's Journal* on 23 June 1899.

of Tintern reflected the state's 'feeble' forestry policy—the lot included 3,000 acres of mature woodland—though it is likely that the importance of the abbey helped the process along; most of the estate was sold off in 1901.[34]

The public debate outside of Westminster focused on the possibility that the National Trust might raise the money for the purchase: it advocated the purchase of private property to preserve public access. This development was part of a longer historical process whereby largely voluntarist bodies had been established to secure public access to 'open spaces'. The initial impetus had been the threat to common land by agricultural improvement, but, from the 1860s, developmental pressure from housing increasingly began to tell. The suburbanization and privatization of the countryside in areas within commutable distance of large cities seemed to pose the most immediate threat, but new kinds of bourgeois prosperity had extended this threat further afield. Around the time Muckross came onto the market, the National Trust had already begun its defensive purchases in the Kent and Surrey hills just south of London and in locations further afield, most obviously in the Lake District. These small parcels of land, much smaller than the Muckross demesne, made the importance attached to the fate of the Muckross demesne all the more striking.

That said, Killarney was not quite the Trust's first foray into Ireland. That same year, the Dowager Duchess of Egmont had presented Kanturk Castle in County Cork to the Trust. This 'curious relic of Elizabethan Ireland' received a somewhat fanciful write-up in the Trust's annual report. It was said to be 'raised by a savage chieftain named Donough McCarthy, of whom tradition says he stopped all wayfarers and compelled them to labour at his castle'; it was to have a glass roof, until the local English settlers grew suspicious of McCarthy's intentions and successfully appealed to the Privy Council to stop the work just as the battlements were about to be raised. It would never be finished and never inhabited.[35] The Trust retained Kanturk Castle as its sole Irish property until accepting the recommendation in 1969 that it should be offered to the Irish government's Ancient Monuments Board.[36]

The influence of Irish peers within the Trust was clearly a factor. At a meeting at Grosvenor House, hosted by the Duke of Westminster, the Trust declared its interest in acquiring 'Killarney as a great national park after the manner of the Yellowstone region in America'.[37] National Trust luminaries like George Shaw Lefebvre, James Bryce, and the Irish peer the Marquis of Dufferin were joined by the Earl of Cork, the Earl of Mayo, and Lord Rosse,

34 J. Winter, *Secure from Rash Assault: Sustaining the Victorian Environment* (Oakland, CA, 1999), p. 98; J. Davies, 'The End of the Great Estates and the Rise of Freehold Farming in Wales', *Welsh History Review*, xvii (Dec. 1974), p. 190; *The Times*, 29 Sept. 1900.

35 National Trust Report, 1899–1900, pp. 7–8.

36 National Trust Report, 1968–9, p. 49.

37 *Freeman's Journal*, 23 June 1899.

as well as the Mayor of Dublin. Letters of support came from the Earl of Meath, the Earl of Carlisle, Joseph H. Choate, the American Ambassador to Britain, and W. E. H. Lecky, historian and MP for Trinity College Dublin. Incidentally, the patron of the National Trust was Princess Louise, wife of the Duke of Connaught, Queen Victoria's seventh child and third son, for whom the purchase of Muckross demesne had been mooted on the eve of their wedding.[38] Octavia Hill, one of the three founding members of the Trust, appears not to have taken an interest in the proposed purchase.[39]

Dufferin, a founding member of the Trust, proposed the motion that it was 'deeply to be deplored' that the Lakes of Killarney, 'perhaps more widely celebrated for their natural beauty than any other spot in the British Isles'—that claim again—might be closed against the public or 'marred by vulgarising and incongruous treatment'. He compared their efforts to the success of the US and Canadian governments to protect Niagara 'from an innumerable number of sheds and shanties': Muckross needed to be protected from the 'jerry builder', the enemy of all preservationists. The *Irish Builder*, similarly alarmed, supported the Trust, believing Killarney should not be made 'a site for "eligible villas"'.[40] The *British Architect*'s wry account of the meeting—'most impecunious lot'—quoted Dufferin to the effect that as an Irishman his instincts 'would have led him to appeal to the generosity of a benevolent Government'.[41] Lecky tapped into a broader preservationist and open spaces concern, writing indignantly that he hoped Muckross would remain 'free from hideous advertisements, which degrade and disfigure some of the most [celebrated] scenery in England', suggesting this would have the full support of Irish public opinion.[42] Official Trust reports identified the threat as coming from 'American millionaires' or 'speculators who would exploit the beauties of the district for commercial purposes'.[43] The *Irish Independent*'s report of the meeting deployed a sharper language still, making arguments as cosmopolitan and anti-plutocrat as they were nationalist. The Trust's aim was 'to save Killarney from the hands of the money-grubber and to preserve its beauties, untrammelled by commercialism, for all citizens of the world'.[44] Given the *Independent*'s Parnellite

38 *London Journal*, 27 July 1878, pp. 53–4.

39 No mention of Killarney appears in Hill's annual letter to her 'fellow-workers'. See R. Whelan, ed., *Octavia Hill's Letters to Fellow-Workers, 1872–1911* (London, 2005).

40 Quoted in *The Builder*, 15 July 1899.

41 *British Architect*, 30 June 1899.

42 *Freeman's Journal*, 23 June 1899; *Irish Independent*, 23 June 1899; Paul Readman, 'Landscape Preservation, "Advertising Disfigurement" and English National Identity, c.1890–1914', *Rural History: Economy, Society, Culture*, xii (2001), pp. 61–83.

43 National Trust Report, 1898–9, p. 16; National Trust Report, 1899–1900, p. 12.

44 *Irish Independent*, 23 June 1899.

politics, it is striking how much faith it had in the 'character of the persons who comprised the meeting and the earnest nature of their remarks', though it also reminded its readers that this was a nationalist question, repeating the claim that an 'Irish Parliament would at once grant a sum to purchase it for the Irish nation'.[45]

By the late 1890s, the Trust had established a clear modus operandi in cases like this. It sought an option to purchase at an agreed price, agree a deadline, and then raise the money through a public appeal. This had worked on several previous occasions, though the method generally relied on the vendor being willing to sell below market value. The Trust's proposed price, £40,000, was far in excess of any previous appeal but the vendor's solicitors maintained this was half what they were looking for. With purchase therefore discarded as a feasible option, the Trust sought agreement on public access. Sir Robert Hunter, another of the Trust's three founding members, visited the site and on his return suggested that the Trust should seek to acquire from the purchaser the freehold to Muckross Abbey, Dinish Island, Old Weir Bridge, Lord Brandon's Cottage, and Tore Waterfall, all iconic Killarney sites.[46] Alternatively, the Trust was ready to enter into partnership with a potential purchaser willing to 'make an arrangement in the interests of the public'. By taking responsibility for regulating public access, the Trust would formalize 'privileges' the public had 'enjoyed for many years', thereby enhancing rather than diminishing 'the amenities of Muckross'. The offer would effectively make the Trust manager of the estate, resolving 'the anomaly of the absence of legal access to a national possession of such world-wide fame as the Lakes of Killarney' and allowing a 'man of wealth' to acquire a 'property of rare beauty' and confer an 'inestimable boon upon the nation'.[47]

Considerable excitement accompanied the sale itself.[48] Conducted in the Ancient Concert Rooms in Dublin, James H. North, the well-known and flamboyant auctioneer, took the stand. He extolled not just the qualities of the

45 Ibid.

46 National Trust Report, 1899–1900, p. 13.

47 Hugh Blackiston, Secretary to the Trust, writing to *The Times*, 2 Nov. 1899.

48 A full description of the lot was offered by the *Irish Builder* (1 June 1899): 'The sale, by private treaty, is announced of the far famed Muckross Estate, including Muckross House Demesne (which entirely surrounds the middle Lake of Killarney), deer forests, game reserves, fisheries, lakes mountains, and islands, containing altogether 13,000 acres. The tenure is fee-simple save as to 227 acres, which are held for ever subject to a head rent. In the sale will be included Muckross Abbey, Torc Waterfall, the Devil's Punchbowl (part of), O'Sullivan's Cascade, Dinish Island, Brickern Islands, Lord Brandon's Cottage, Mangerton Mountain (part of), the Old Weir Bridge (part of), the Meeting of the Waters, &c., in fact the finest portion of the scenery of Killarney. The deer forests are the best in Ireland, and the cock coverts are probably the same'.

product, but also its wider importance to the Irish nation. The degree to which North presented the estate as a form of common property seems remarkable; presumably, the Herberts required it be presented in this way. This, the 'most historic auction ever held in Ireland', included 'all that is best of our great National asset in the country's scenic resources—the world-famed Lakes of Killarney'. North suggested that, until now, 'We were almost unaware of any private ownership, and while the world might claim the right to see and admire, the treasure was ours, and ours only. The earth has only one Killarney and we possess it'.[49] This was nationalism as a form of possessory right. North continued: This 'beauty', the 'birthright' of the Irish people, culminates 'in the mountains, lakes, glens, woods, and waterfall of Killarney, over which sunshine and shadow for ever sweep in fitful alternation, bewildering the beholder at every new revelation of [un]suspected loveliness'. Whether these remarks discouraged bidding is hard to say, but there was no response to North's opening at £80,000. The starting price was dropped to £35,000 but the bidding only reached £50,000, at which point North brought proceedings to a halt.[50] The sale was a flop.

A week later the estate had been sold by private treaty for £60,000 to Arthur Guinness, the first Lord Ardilaun. This was widely considered good news. Although a staunch tory and unionist, Ardilaun's reputation as a philanthropist was well-founded, not least on his decision to landscape St Stephens Green and present it to the Board of Works for the use of the citizens of Dublin. The *Irish Independent* waxed obsequious, describing the 'senior member of the Guinness family' as 'too large hearted in his practical philanthropy, lavished in his own country, to lay himself open to any imputation of having purchased Killarney for the sordid and unpatriotic purpose of revoking or modifying by a tittle the prescriptive right of the Irish people to glory in Killarney as a magnificent scenic asset of the nation available to all comers'.[51] Note that the legalistic tone of this observation—'prescriptive right'—had no legal status. As such, the emergency might have passed, but the Trust was not content that the protection of the public interest relied upon the benevolence of the owner.[52]

Ireland, then, cannot boast of the UK's first national park, but it is clear the story did not end in 1899.

49 *Irish Daily Independent*, 22 Nov. 1899.
50 Ibid.
51 *Irish Daily Independent*, 27 Nov. 1899.
52 National Trust Report, 1899–1900, p. 14.

*

Ardilaun was not much interested in his Killarney estate. His maintained the toll regime and let the house on a seasonal basis as a hunting and fishing lodge. In 1910, William Bowers Bourn, a Californian businessman and garden enthusiast, took the lease and in August 1911 bought the estate for £50,000. His daughter, Maud, married Arthur Rose Vincent, a barrister on the Munster circuit who had joined the Foreign Office judicial service. On 18 August 1916, Bourn made the estate over to a trust but granted right of property to Vincent. The indenture included a remarkable clause, albeit one virtually incomprehensible on first reading:

> In default of and subject to any and every such appointment to the use of the Trustees in fee simple Upon Trust to convey the same to the King's Most Excellent Majesty in fee simple or as he may direct to the intent that the same may if His Majesty thinks fit so to ordain be constituted a Royal Park or be dedicated as and for a National or Public Park or for such other purposes as to His Majesty may seem meet.[53]

Sixteen years later, Vincent, now a senator in the Irish Free State, wrote to President de Valera about his desire to make the estate over to the state.[54] As Vincent explained, since Bourn acquired the estate from Ardilaun it had been greatly 'improved', including the significant development of its forestry operation. 'It is now', he argued, 'in what one might call perfect condition', but he felt that the 13,000 acres was too big an undertaking for an individual.[55] De Valera turned the matter over to the Attorney General and the Minister of Finance, the former visiting the estate shortly afterwards.[56] He returned to Dublin impressed and his favourable account of Vincent's improvements provides important insight into the historical production of this landscape. Vincent had cleared the wood and scrub that stretched from Muckross Lake to the house, allowing for an unobstructed view of the mountains and lake; he had also cleared away a substantial volume of soil to reveal a natural rock garden some three-and-a-half acres in scope, altogether creating gardens and lawns that although 'artificial' were in 'complete harmony with the natural beauty around'. A striking claim that revealed the tension between intrinsic and produced charismatic landscapes.

53 Registered in Registry of Deeds, 29 Aug. 1916.
54 Arthur Vincent to de Valera, 28 July 1932 (TAOIS/S 6355 A).
55 Ibid.
56 De Valera to Vincent, 29 July 1932 (TAOIS/S 6355 A).

Most impressive, however, was Vincent's forestry operation. He had clear-felled hundreds of acres of land, wired it off from deer, rid it of rabbits by trapping, and had it replanted. As the attorney general observed, 'It is poor land and the timber grows slowly but its hardness makes it superior to any other timber grown in Ireland', a quality recognized by Great Southern Railways, who used the wood for their points. Again, Vincent's aesthetic achievement was emphasized. Clearing the understorey from the woodland had opened up views of the lake and mountains, creating an exceptional asset of potentially high public value: 'it hardly needs another touch to make this estate the finest public park in the world'. At present, the park was only visible to the public as they drove through, but the attorney general was sure the state could easily open the park to the public, radically extending their access.[57] The farm, run on 'scientific' lines, was impressive also, another testimony to Vincent's modernizing and improving outlook, but most striking was Vincent's 'handling of the labour problem'. His sixty-five employees were all locals, had access to a clubhouse, reading room, and subsidised produce, allowing Vincent to claim that none of his employees or their families was malnourished. Tellingly, Vincent boasted that after an absence of three years, he had arrived unannounced to find his employees properly engaged in their work. Careful bookkeeping and a business-like approach helped account for his success, though the estate nonetheless required an annual subvention of £11,000: £3,000 was raised by the house, £3,000 from the toll roads and fisheries, and Vincent made up the remainder.[58] A benign, if absent, landlord was expensive.

On 8 October 1932, a memorandum of counsel was drawn up between Vincent and the Minister for Finance. The estate would be gifted to the Minister of Finance, and the Commission for Public Works would be entrusted with its management and control. The park would be named 'The Bourn-Vincent Memorial Park' in memory of Maud Bourn-Vincent, Vincent's daughter, who had died in New York in February 1929. Vincent retained the right to erect a monument in her memory, which the state would be obliged to maintain, and the estate was to be taken on as a going concern, in which the services of its gardeners, foresters, gamekeepers, gatekeepers, and workmen would be retained at the going rate and their right to residence on the estate guaranteed, a condition that stored up trouble for the future. Vincent insisted that he should incur no costs in relation to the transfer of ownership, making him exempt from stamp duty or death duty should he die within three years, and that the transaction be complete by 31 December 1932.[59]

57 SR 39/22, 2005/77/34 Office of the Attorney General.

58 Ibid.

59 See copy of Memorandum of Counsel, Henry Maloney, 8 Oct. 1932 (TAOIS/S 6355 A).

On 28 October 1932, the Cabinet decided to accept the gift. The government announced that the estate and house would pass into state ownership as a 'free gift to the Nation for the purposes of a great National Park'. The political significance of the transfer in ownership was foregrounded: 'The beauties of Killarney are world-famed. Henceforth they can by no mischance become the monopoly of a privileged few: they are part of the heritage of the nation'.[60] The *Irish Times* was on message. In a lyrical editorial, it outlined the social and political significance of Vincent's 'fine act of faith in the country's future' and his 'hope that peace and good-will may spread from this noble playground of the State'. The presence on the estate of the ruins of Muckross Abbey helped shape how the gift was received. The *Irish Times* described the exceptional beauty of this 'catholic haven of rest, health and pleasure' and declared that henceforth 'the Free State will boast a national park superior, within its perfect limitations, to anything England or American can show'.[61] Much the same was heard during the rapid, two-day progress of the bill through the Dáil.[62] Certain of de Valera's opponents were at their most patrician. Cosgrave, leader of the opposition, was extraordinarily deferential, recognizing the generosity of this 'princely' gift and 'the remarkable evidence of citizenship on the part of a princely family', suggesting it might be 'graceful' to invite Vincent to occupy the house as he wished and for the Bourns to be granted 'freedom to visit'.[63] Other agendas were at work too. Frank MacDermot, independent TD for Roscommon, explained that Vincent had asked him to explain

> that he would wish the gift to be regarded as symbolic of the desire of himself and his class for appeasement in this country and for good feeling between all the classes that compose our State and as symbolic of the desire of every section of our community to be regarded as welded solidly into one homogenous whole and to be associated with the most ardent aspirations for the prosperity and glory of the Irish people in the future.[64]

Much the same was heard from across the political spectrum in the Seanad. In a remarkable speech, Thomas Westropp Bennett, the Cathaoirleach, observed that Muckross might have been the inspiration for 'The Princess', Tennyson's 'finest lyric', but its significance was not to be found in this English example.

60 Copy in SR 39/22, 2005/77/34 Office of the Attorney General.
61 *Irish Times*, 24 Nov. 1932.
62 Dáil Éireann, Parliamentary Debates, xlv (Nov.–Dec 1932, 6 Dec. 1932), cols. 776–777.
63 Dáil Éireann, Parliamentary Debates, xlv (7 Dec. 1932), col. 947.
64 Dáil Éireann, Parliamentary Debates, xlv (7 Dec. 1932), col. 948.

> Muckross means more than that, however, to the Gael, for in the ruined priory of its age-old abbey is the last retiring place of the noblest warriors of the great clans of the McCarthys, the O'Sullivans and the O'Donoghues. Here, too, in the chancel, lie two of the greatest of the Munster poets, Egan O'Rahilly and Own Rua O'Sullivan, the Robert Burns of the native Irish. As the knowledge of our mournful but proud history becomes more widely disseminated, as it will from year to year, our Irish people will view this 'quiet, watered land,' which is now about to become their property, with the deepening affection which comes from knowledge; and when memories of old rancours, now dying, are long dead, our children's children will regard the gift, not as an act of restoration from Gall to Gael, but as a bestowal of part of our national heritage, which had at different times for centuries been in the possession of both Gall and Gael, upon a united people in which both elements are happily blended and which is informed with the virtues of both.[65]

On 15 December 1932, the final stages of the bill were passed, preparing the legal ground for the state's acceptance of the gift a fortnight or so later.[66]

*

Kevin James observes that US National Parks 'created vast stages for the performance of citizenship through bodily engagement with iconic national landscapes'.[67] This was emphatically not the case in Ireland. The acquisition of the estate and house proved a burden to the Office of Public Works. The *Irish Times* periodically drew the public's attention to the government's apparent failure to promote the park's exceptional qualities. Now the Irish state was found wanting. Had such a gift been made to the government of Germany, England, Italy, or the USA, the newspaper complained, the 'nation would have rung with the news. Its beauty would have been made known by posters and films in every public place', making it a destination for hikers, cyclists, canoeists, and motorists. Particularly lamentable was that visitors tended to be older and English or American.[68] Muckross was 'a National Park only in name' and would only fulfil its potential when the people came to it. It must be advertised, it must be accessible, it needed trails, its shooting

65 Seanad Éireann, Parliamentary Debates, xvi (13 Dec. 1932), col. 409.
66 Seanad Éireann, Parliamentary Debates, xvi (15 Dec. 1932), col. 607.
67 James, *Tourism, Land and Landscape*, p. 137.
68 *Irish Times*, 15 June 1936.

rights must not be sold, for this was off-putting to visitors, and the house must not be leased but converted into a youth hostel.[69] The deer-stalking and shooting rights were discreetly advertised elsewhere in the newspaper. According to a moralizing Swiss reader, the creation of the Swiss National Park, largely held by the Swiss government on a ninety-nine-year lease, had eliminated all personal advantage and material profit from the land. 'That alone is a piece of national education', the writer solemnly opined, 'a patriotic accomplishment making for unity and purity, the value of which cannot be too highly estimated'.[70]

In 1943, the Office of Public Works prepared a memorandum in response to the latest 'direct or oblique attack' by the *Irish Times*, which had made the case for a 'semi-State' body analogous to the National Trust to accept and manage property on behalf of the people.[71] It is a fascinating document that offers surprisingly rich insight into government thinking. Outlining the problems incurred by a gift that, in practice, was 'a definite liability', the Office explained how, additional to subsidies paid to maintain the farm, the estate had already cost the state some £30,000. The most difficult questions concerned the house, which was unoccupied but for the caretaker and housekeeper. As late as 1938, Vincent alluded to his hope it might become a youth hostel, but an An Óige inspection of April 1935 concluded that adapting its splendid interiors for dormitory accommodation at a shilling or two a night was impossible. Such concerns, however, were nothing to the moral challenge imposed by the building: 'the Commissioners do not consider that they would be able to exercise sufficient or satisfactory control over the young people of both sexes who might be expected to lodge in the house if it were converted into a hostel'.

Repeated suggestions that the house be leased as a hotel were always met with the response that Killarney already had plenty while the possibility that it be the country residence of the president also came with a warning: 'it is doubtful if anyone would wish to spend very long periods there as the district is generally considered relaxing and somewhat enervating'. Its prospects as a convalescence home had been dismissed in 1940 on the grounds that Kerry's climate was less than ideal and the cost of maintaining the building would be too high; letting to a tenant was a possibility, but only if this did not inhibit public access and the tenant would 'have to refrain from the use of the house for gaming or casino purposes'; the idea that the Department of Agriculture should use it for a college, which the prize cattle made attractive, was undermined by the fact that of the 11,000-acre estate only about 300

69 *Irish Times*, 16 June 1936; see also 20 Mar. 1937 and 11 Nov. 1938.
70 *Irish Times*, 1 July 1936.
71 *Irish Times*, 13 July 1943.

acres were arable grazing and meadow lands. To grant use of the building as a retreat to a religious order was thought impolitic given how much state property was already in the hands of the religious orders and, in any case, the proximity of the rock garden to the house would raise questions of privacy at odds with the original intentions of the donor. This left the Office's preferred option of a horticultural school under the aegis of the Department of Agriculture, a suggestion which strongly echoed the significance the Attorney General had attached to Vincent's forestry operation more a decade earlier.[72]

The debate rumbled on, the tension between the intentions of the donor, National Park ideals, and the impecuniousness of the state remaining unresolved. Only in the early 1960s did the government finally began to give serious consideration to how to manage and resource the house and the estate as a public amenity, but so opaque was the government's strategy that opposition deputies could still pointedly ask ministers if the Bourn Vincent Memorial Park was a National Park at all—a question that carried an additional charge now that England and Wales had firmly established National Parks. Only in the 1970s was a serious commitment made to manage the National Park according to recognizable international standards, though to date that endeavour has been dogged at every stage by the invasive success of Rhododendrons.[73] This is another story, which must form a part of any account of the modernization of the Irish state in the Lemass years and beyond. More immediately reflective of the history of Ireland's first national park and, indeed, the degree to which the Irish government adopted the outlook of its predecessor, was the fate of the neighbouring Kenmare Estate.

Following the death of Lord Kenmare in 1952, the estate was inherited by his niece Beatrice Grosvenor, who decided to sell to meet death duties. The concerns of the 1890s were immediately revived. De Valera insisted Liam Cosgrave, the Taoiseach, explain how the government would prevent purchase by 'a foreign syndicate' or persons who might exploit the estate 'in a manner detrimental to the national interest'; the Killarney Tourist Association launched a fund to purchase the estate for 'Killarney and the nation'; Sinn Féin suggested descendants of evicted tenants should receive the land as compensation; and Bord Fáilte suggested it should be purchased 'on a national basis' and added to Bourn Vincent Memorial Park. The government refused, saying to intervene would set a dangerous precedent.[74]

72 See TAOIS/S 6355 A.

73 For an overview, see Fogarty, *Whittled Away* (Cork, 2017), pp. 91–3.

74 Dáil Éireann, Parliamentary Debates, clviii (14 June 1956), cols. 398–399; Eric Zuelow, *Making Ireland Irish: Tourism and National Identity after the Civil War* (Syracuse, NY, 2009), pp. 186–7; *The Times*, 30 June 1956.

In the event, a private investor took up the opportunity. J. Stuart Robertson, a Florida real estate broker of mixed Scottish and Irish background, secured the purchase in August. Defensive ministers clung to his commitment not to harm Killarney's traditions or its natural beauty. Much was made of how this protected the rights of the boatmen and jarveys to ply their trade.[75] Misgivings soon arose. Robertson's 'share Killarney plan', by which fifty summer residences would be built on Ross Island for wealthy Americans at a cost of £250,000, did not go down well in all quarters.[76] In the Dáil, Jack McQuillian, independent TD for Roscommon, opposed the plan in vociferous terms. Not only had the estate been sold to 'people outside this State', but, he asked, was 'it not a fact that the present occupier—and I use the word "occupier"—intends without doubt to establish a rest camp in [the] Kenmare estate for the purpose of facilitating a bunch of neurotic Hollywood personalities in between their matrimonial swopping activities?'[77] McQuillian's refusal to withdraw his remarks led to his suspension from the chamber, rendering his intervention a disruption rather than a challenge to conventional, free market thinking. Soon after, Robertson sold some of his shares and formed a new company, Killarney Estates Limited, comprising eight Irish-American proprietary members and two Irish-American Catholic priests.[78] The company aimed to build twenty detached homes for American millionaires. The fate of this plan is unclear, but Robertson sold his interest in early 1958, and the new chair, Ignatius Aloysius O'Shaughnessy, Irish-American president of the Globe Oil Company, stated that they no longer planned to build on Ross Island and henceforth the estate would be used only for holidays by the twelve shareholders.[79] In December 1960, John McShain, building contractor and Catholic philanthropist—the 'Man who built Washington'—became the sole owner.[80] In 1972, he presented Innisfallen Island to the nation and placed Ross Castle into the guardianship of the Board of Works. The remainder of the estate was later sold to the state at below market value; the state took full possession in 1998 following the death of McShain's widow. After 250 years, the ancient dispute between the Herbert and Kenmare estates over the bed, soil, and fishing rights of Upper Lake and the Long Range was finally resolved, and Killarney National Park as we know it today came into being.

In contrast to Britain, where the National Park Authorities are planning authorities with additional powers, Ireland has National Parks on a broadly

75 *The Times*, 11 Aug. 1956.
76 *The Times*, 21 Nov. 1956.
77 Dáil Éireann, Parliamentary Debates, clx (5 Dec. 1956), col. 1835; *The Times*, 6 Dec. 1956.
78 *The Times*, 15 May 1957.
79 *The Times*, 14 Aug. 1958.
80 *The Times*, 29 Dec. 1960.

North American model, albeit on a vastly smaller scale and without the same ecological value. Most land in British National Parks is privately owned—there are no National Parks in Northern Ireland—whereas National Parks in the Republic are owned and managed by the state. With respect to Killarney, this is as much a question of luck as anything else. Contrary to the fears of early preservationists, its status has depended on a complex mix of patrician benevolence, American plutocracy, and the declining viability of the remnants of two great landed estates; the Irish state effectively behaved like the Natural Trust, relieving one owner of a costly burden and making an apparently advantageous purchase from the other. The park boundaries are not determined by the logics of natural or scenic characteristics and instead archive the demesne lands of two Big Houses, but the fact that these *were* demesne lands reflects a continuity in sensibility that stretches from the eighteenth century through to the present day.

Much the same can also be said of Glenveagh National Park in County Donegal. It was acquired by the state in the mid-1970s from Henry Plummer McIlkenny (1910–86), an Irish-American art collector who bought it in 1937. What appealed to McIlkenny is doubtless what appeals to visitors today, an effect achieved by John Adair's infamous eviction of forty-seven families and the levelling of twenty-eight homes in 1861. As W. E. Vaughan observes, 'one of the few pieces of the land of Ireland to pass into the ownership of the people of Ireland does so because of a great act of landlord tyranny in 1861'.[81] Here, then, is another point of comparison with the US national parks system. The US parks might be based on notions of wilderness, and Glenveagh on nineteenth-century scenic ideals, but both stem from violent acts of appropriation. The US park authorities expelled indigenous populations and later arrivals with a ruthlessness that mirrors Adair.[82]

Killarney does not boast quite such an unpleasant history. The Herberts and Kenmares were relatively benign landlords and, from this perspective, Killarney's notorious unruliness, that lower-class entrepreneurialism, starts to look like relatively light-touch estate management. It should no more be assumed that the National Trust or the UK government would have been quite so tolerant as there is any guarantee that a private owner would not restrict access. Despite this, in popular culture there remained a strong sense, so clearly articulated in the 1890s, that the public retained rights to Killarney that transcended the rights of private property. In a song of 1937 made famous

81 W. E. Vaughan, *Sin, Sheep and Scotsmen: John George Adair and the Derryveagh Evictions, 1861* (Belfast, 1983).

82 M. D. Spence, *Dispossessing the Wilderness: Indian Removal and the Making of the National Parks* (Oxford, 1999); K. Jacoby, *Crimes against Nature: Squatters, Poachers, Thieves, and the Hidden History of American Conservation* (Berkeley, CA, 2003).

by Bing Crosby, the response to an American visitor asking, 'How can you buy Killarney?' was that the beauties of Killarney, freely bestowed by nature, can be no more bought than the 'stars in the skies', 'two blue Irish eyes', and 'a fond mother's sighs'—not to mention 'the emerald, the shamrock, the Blarney'. And yet, the pre-history of Killarney National Park makes it clear that no amount of anti-materialist transcendentalism can disguise the historical fact that it was quite possible to buy Killarney if you were very rich.

# 7
# Towards an Environmental History of Nineteenth-Century Dublin

*Juliana Adelman*

What would an environmental history of Dublin look like? Why would we need such a history? Environmental history, which has its roots in the American conservation movement of the 1960s, originally focused its attention on rural landscapes and the frontiers of human settlement. Now that few places on earth could be considered untouched by human activity, a more expansive idea of environmental history has emerged to encompass a wide variety of interactions between the human and non-human worlds. Urban environmental historians have seen cities not as exclusively human places, but spaces in which humans are in constant negotiation with the non-human world.[1] The following pages begin to map some directions for an environmental history of nineteenth-century Dublin. First, I show how we might examine the intersection between environmental factors and socio-economic change to understand the distribution of people and businesses in the city. Second, I look at animal businesses in detail to see how changes to the use of animals in the city both reflected and shaped larger processes of change in Dublin over the Victorian period.

## Urban Landscapes

The nineteenth century was a transitional period for the shape and size of most cities in Europe and America. Industrial cities expanded rapidly in area, population, and infrastructure. Manchester, Leeds, and Glasgow are often cited for their dramatic and, to contemporary observers, unnnerving

1 W. Cronon, *Nature's Metropolis: Chicago and the Great West* (New York, 1991); F. L. Brown, *The City Is More than Human: An Animal History of Seattle* (Seattle, WA, 2016).

population explosions.[2] The majority of these cities' new residents were poor, unskilled workers and their arrival coincided with the increasing flight from the city centre of the middle and upper classes who sought the healthier, cleaner environments of the suburbs. Movements of population created rings of differing social and economic status around the city centre, leaving the inner core increasingly impoverished and often polluted. Despite the interest of municipal authorities in controlling nuisances from industries and noxious trades, many factories remained in the city centre throughout the century.[3] Among the trades that produced unpleasant smells and effluvia were a wide variety of animal businesses, including slaughterhouses, dairies, bacon factories, bone boilers, glue manufactories, tanners, skinners, and tallow chandlers. Thus the urban poor and the city's non-human residents were often left to share the same insalubrious space.

Despite Dublin's failure to develop a manufacturing economy, the city shared the socio-economic and geographical transitions that marked many industrializing cities. The establishment of an outer ring of wealthy suburbs meant that, in some cities, the middle and upper classes could largely avoid the sights, sounds, and smells of the poor quarters and move between their exclusive neighbourhoods and urban commercial districts. The arrival of relatively inexpensive tramlines threatened this, enabling the working classes also to commute from the periphery. Dublin, too, experienced social segregation but of a distinct form. Suburban development accelerated in the 1860s, unaided as yet by trams. By the opening of the first tramline in 1871, the south-eastern fringes and the coastal areas, the latter of which had previously been holiday resorts, were well-established suburbs for urban professionals. They were also largely Protestant and politically conservative in their outlook, and determined to remain administratively separate from the city itself, which they viewed as corrupt and hostile.[4] Many reformers advocated the inclusion of the suburbs into the municipality as a way of raising revenue to tackle urban blight. Sir Charles Cameron, the city's first public analyst, resided for his entire career in the salubrious suburban township of Pembroke.[5] Cameron believed that the cause of public health would be aided by the inclusion of the suburbs in the city.

Parliament and the Dublin Corporation attacked the city's problems with many rounds of legislation aimed at making the city cleaner and more healthful.

2 T. Hunt, *Building Jerusalem: The Rise and Fall of the Victorian City* (London, 2005), p. 18.

3 P. Knox and S. Pinch, *Urban Social Geography: An Introduction* (Harlow, 2006), p. 31.

4 S. Ó Maitiú, *Dublin's Suburban Towns, 1834–1930: Governing Clontarf, Drumcondra, Dalkey, Killiney, Kilmainham, Pembroke, Kingstown, Blackrock, Rathmines, and Rathgar* (Dublin, 2003).

5 L. Carroll, *In the Fever King's Preserves: Sir Charles Cameron and the Dublin Slums* (Dublin, 2011).

The nineteenth century marked a significant shift in who was to maintain and improve the urban environment. After the Municipal Corporations Act 1840 and the Dublin Improvement Act 1849, local government was both more representative and more responsible. The reforms had obvious political and social effects. A small electorate of only around 7,000 persons (who elected thirty-five councillors and fifteen aldermen) prevented an immediate takeover of city politics by Catholics, but the city council became increasingly Catholic and nationalist over the century. The reforms also had environmental impacts. Dublin Corporation took over a wide range of services that had often been controlled by separate government agencies. These included the regulation of trades that produced pollution, paving and lighting of streets, setting housing standards, licensing and regulating horse-drawn traffic, widening of streets, and creating and maintaining infrastructure such as water, drainage, and bridges. Many animals and animal businesses came under their remit and Corporation employees removed manure from streets, detected unsound and adulterated foods, and carried out regular inspections of manure yards, markets, and slaughterhouses. Unfortunately, the reformed corporation took a similar attitude as the old towards anything that required the raising of rates, and few reforms could be effective without significant investment.

The reformed corporation inherited a city with a number of socio-economic gradients that had been established before Queen Victoria's reign and which it had not altered by the end of the century. Historians and geographers have grappled with the meaning of the socio-economic and cultural distribution of people within the urban landscape. As recognized by any urban resident today, the city shows an uneven distribution of wealth, services, and population. While the distribution can be mapped, providing explanations for its occurrence is more difficult.[6] Environmental factors have been increasingly implicated in the inequalities displayed by contemporary cities and such factors have been observed in action in the past.[7] Animals and animal industries, for example, both affected and reflected the status of particular urban neighbourhoods.[8]

6 Knox and Pinch, *Urban Social Geography*, pp. 21–31.

7 For example, K. Cruikshank and N. B. Bouchier, 'Blighted Areas and Obnoxious Industries: Constructing Environmental Inequality on an Industrial Waterfront, Hamilton, Ontario, 1890–1960', *Environmental History*, ix (2004), pp. 464–96; M. Klingle, 'Changing Spaces: Nature, Property, and Power in Seattle, 1880–1945', *Journal of Urban History*, xxxii (2006), pp. 197–230; C. Philo, 'Animals, Geography and the City: Notes on Inclusions and Exclusions', *Environment and Planning D: Society and Space*, xiii (1995), pp. 655–81.

8 See, for example, K. Anderson, '"The Beast Within": Race, Humanity and Animality', *Environment and Planning D: Society and Space*, xviii (2000), pp. 301–20; C. Otter,

The factors to be considered in the socio-spatial arrangement of nineteenth-century Dublin are environmental, economic, and cultural. The city is bisected by the River Liffey, which divides it into north and south. The river provides a channel between the countryside on the city's western edge and the sea at its east. Throughout the century, advocates of sanitary reform noted that the Liffey might have provided a channel of fresh and health-giving air. Unfortunately, the river also provided the service of a sewer in the absence of a main drainage system and the air above it was therefore considered a miasmatic stink. As one writer put it: 'The people make offerings to it of cats and dogs and doglings (for the most part defunct), and boots and ashes, and a whole lot of things else; in return for which the grateful water gives forth gases and perfumes'.[9] Most rivers in nineteenth-century cities were treated in the same manner: the 'great stink' of London in 1858 saw suggestions to close the new House of Commons until the smell of rotting sewage from the Thames could be brought under control.[10] In Dublin, the river was also a social and economic barrier. In the eighteenth century, it disconnected the prosperous northern squares from the remainder of the city. In the nineteenth century, the trend was reversed, and the northern parts of the city were left to crumble as wealth became concentrated in the south-east and adjoining suburbs.

Nineteenth-century Dublin also exhibited an east–west divide that baffled those interested in sanitation and health. Sackville Street (now O'Connell Street) on the north side and Carlisle Bridge (now O'Connell Bridge) connected the sites of British and Protestant Ascendancy power, all located in the east but on each side of the river: the residential squares of the Gardiner Estate, the Rotunda Hospital, the General Post Office, the former Irish parliament, and Trinity College.[11] The western portion of the city is situated on an elevated plain and the prevailing winds carry fresh air in from the rural hinterland and pollution out towards the sea. Nonetheless, the western districts remained poor and the wealthy flocked to the eastern portion of the city. This makes Dublin quite different from other cities, where the waterfront was reserved for poor inhabitants and large-scale industry.[12] London's wealth gradient, for example, went in the opposite direction, with the salubrious west (further away from the mouth of the Thames) becoming more and more economically remote from the east as

'The Vital City: Public Analysis, Dairies, and Slaughterhouses in Nineteenth-Century Britain', *Cultural Geographies*, xiii (2006), pp. 517–37; C. Philo and C. Wilbert, eds., *Animal Spaces, Beastly Places* (London, 2000).

9 'Guide to Dublin', *Zoz*, 28 Sept. 1878.

10 Hunt, *Building Jerusalem*, pp. 260–2.

11 C. Maxwell, *Dublin under the Georges, 1714–1830* (London, 1936), pp. 65–70.

12 For example, Klingle, 'Changing Spaces'.

the century wore on.[13] The eastern orientation of wealth in Dublin probably reflects both the eastern orientation of important administrative buildings and the history of land ownership in and around the city, where the controlled development of particular estates in its eastern portion ensured the building of expensive, planned tracts of housing for a wealthy clientele.[14] In addition, the numerous small rivers and streams of the western district allowed the development of small industries, many producing bad smells that would have discouraged wealthier citizens from living beside them.[15] The city analyst noted in 1870 that a group of sparrows flying over an open sewer from the gas and chemical works in Dublin 'dropped dead as suddenly as if they had been shot'.[16]

The qualities of the city's landscape contributed to socio-economic divisions, which were then converted into political ones. For example, the commission appointed to draw new voting wards in 1850 accepted the river as a dividing line. Thomas Larcom, under-secretary at Dublin Castle, reported that 'we found the city so unequally distributed, as regards the amount of individual rating, that it appeared at one time altogether impossible to attain the desired equality in allotting the wards'.[17] The rating refers to the value of a property; the aim of the commission was to create voting districts that were roughly equal to one another in both population and rated valuation. Larcom's commission noted that there was an extreme gradient of wealth between east and west, with the average property value in the latter being only £20 versus £100 in the former. Large gradations of value were noted within each district; the focus of wealth in the west was located near the river and in the east near to its grand Georgian squares.[18]

Infrastructural developments often reinforced socio-spatial patterns. The Midland and Great Western Railway line with its terminus at Broadstone in the north-west of the city, for example, was built during the Great Famine with a speculative eye to the cattle trade.[19] The line followed close to droving routes and was to link up with and eventually replaced a series of canals that had served a similar purpose. The Dublin to Kingstown railway, the

13 Hunt, *Building Jerusalem*, p. 390.

14 E. Sheridan-Quantz, 'The Multi-Centred Metropolis: The Social Topography of Eighteenth-Century Dublin', in P. Clark and R. Gillespie, eds., *Two Capitals: London and Dublin, 1500–1840* (London, 2001), pp. 265–96.

15 J. Prunty, *Dublin Slums, 1800–1925: A Study in Urban Geography* (Dublin, 1998), p. 55.

16 C. Cameron, 'On the Chemical Composition and Fertilizing Value of the Sewage of Dublin', *Dublin Builder*, 1 Feb. 1865, p. 4.

17 Papers of Thomas Larcom, NLI MS 7461.

18 Ibid.

19 Michael Killeen, 'Broadstone: Railway Station to Bus Garage', *Dublin Historical Record*, xxxiv (1981), pp. 140–54.

first line in Ireland, acknowledged and cemented the importance of the city's southern suburbs to its middle classes.[20] Railway development could also solidify districts of poverty both by the intrusion of noise and dirt and by the destruction of housing, driving further crowding into tenements.[21] By contrast, railways were more successful adjuncts to change in younger cities experiencing rapid expansion. In Chicago, for example, the cattle market was moved to the western edge of town and linked to a new railway spur. Cattle were brought to market and forwarded by rail (usually as dead meat) without the need to go through the city centre.[22]

The need to share the urban environment with animal bodies (living and dead, for work and food) also affected Dublin's social and economic divisions and how these changed over time. Certain urban animals were particularly associated with poverty and filth while others were more widely distributed. Dublin can be usefully viewed as four quarters, with differing characteristics and different mixtures of animal businesses. The Liffey serves as the north–south dividing marker. As the east–west axis we can use a rough line extending south from Capel Street, skirting Dublin Castle, and continuing down Camden Street (see Figure 7.1). The south-east quarter had both the smallest number and the lowest density of animal businesses, and those businesses were almost entirely related to horses. The north-east initially exhibited a similar pattern, but by 1901 its animal businesses had expanded and diversified substantially, and the area now included the densely packed meat markets of Moore Street and Riddall's Row. The south-west began with a dominance in trades associated with the processing of animal by-products, clustered with its numerous butcher's shops. The by-products trade had diminished and dispersed by 1901, but the area continued to have clusters of butcher's shops. The north-west expanded dramatically with the booming of the cattle market, despite the market's removal to the urban periphery. By 1901, the north-west was dominated by the livestock trade but had a substantial smattering of other animal businesses.

The next section explores this animal geography and how it changed over time. Figure 7.1 provides an overview of the issues discussed below. Using data from Thom's Directories, the map shows the persistent concentrations of particular businesses in areas of the city. Slaughterhouses, for example, could be found in many streets, but I have represented the highest areas of concentration on the map. The vertical black line runs down Sackville Street (now O'Connell Street) on the north side and Westmoreland Street and Grafton

20 John Dunne, 'The First Irish Railway by Shriek and Smoke to Kingstown', *Dublin Historical Record*, xliv (1990), pp. 44–6.

21 Hunt, *Building Jerusalem*, p. 394; Prunty, *Dublin Slums*, p. 52.

22 Cronon, *Nature's Metropolis*, p. 209.

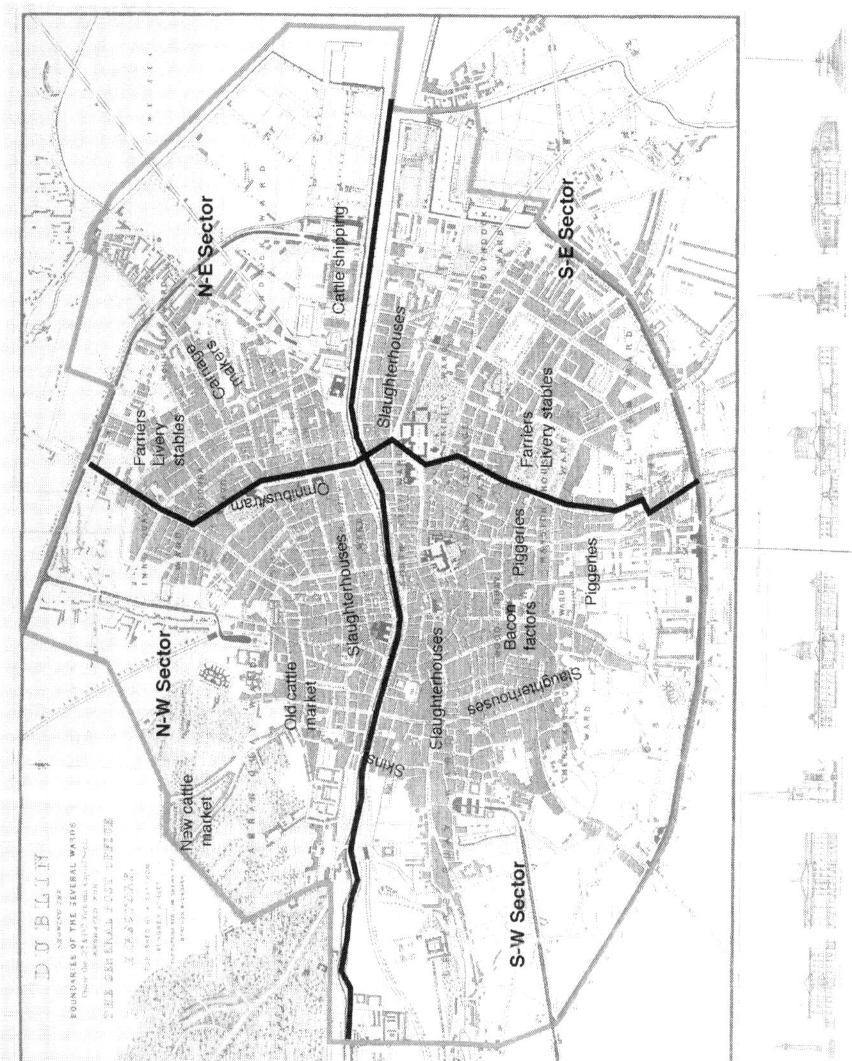


7.1 Map Showing Concentrations of Animal Businesses in Nineteenth-Century Dublin
Data compiled from street directories. The author wishes to acknowledge the assistance of Dick Tobin.

Street on the south side, dividing the city into west and east. The horizontal black line follows the river Liffey and divides north and south. The lines roughly break Dublin into four quadrants with different concentrations of animal businesses.

## Animal Geography

Dublin began the nineteenth century with an arrangement of businesses that, if not medieval, certainly had its basis in the seventeenth and eighteenth centuries. As a consequence, Dublin looks different from those British and American cities often associated with nineteenth-century urban reform. The city's well-established business pattern proved resistant to change. Although many aspects of medieval Dublin have vanished, including buildings and most of the city wall, the contemporary city retains much of the medieval streetscape. Local trade specializations were often related to access to environmental resources, principally water. The area of the city designated as the Liberties (outside the medieval city walls) was converted into an industrial area between the late seventeenth and early nineteenth centuries, making use of the River Poddle. The Liberties remained an area of small industry well into the nineteenth century where the Guinness Brewery, tanners, skinners, and the declining weaving industry were all clustered.[23] The situation of Smithfield Market in the north-west of the city was close to good-quality grazing pastures and droving roads of at least early modern origin.[24] The localization of trades was recognized in the ballad 'Dublin Jack of All Trades', where the singer related his travels around the city performing different types of work: 'In College Green a banker was, and in Smithfield a drover, / In Britain Street a waiter, and in George's Street a glover'.[25] The ballad, perhaps originating in the mid-nineteenth century, related each trade to a specific city street and suggests one of the reasons that business patterns tended to stay the same: they were a part of the familiar landscape of the city, allowing Dubliners to know precisely where to go for particular services and products.

Urban livestock and agricultural markets, most established in the sixteenth and seventeenth centuries, proved particularly intransigent in Dublin as in other cities. London's Smithfield Market, for example, was hemmed in on all sides by dense housing at the start of the nineteenth century. Its situation

23 N. McCullough, *Dublin, an Urban History: The Plan of the City* (Dublin, 2007), p. 58.

24 L. Clare, 'The Rise and Demise of the Dublin Cattle Market, 1863–1973', in D. A. Cronin, J. Gilligan, and K. Holton, eds., *Irish Fairs and Markets: Studies in Local History* (Dublin, 2001), pp. 180–205.

25 Quoted in C. Lowth, 'Dublin Jack of All Trades', *Dublin Historical Record*, lxi (2008), pp. 169–82.

was deemed inappropriate and inconvenient for at least thirty years before the market was actually moved to a more open location.[26] Smithfield in London retained a dead meat market and the site of the new livestock market was soon surrounded by further housing. Dublin's Smithfield Market faced similar problems and it, too, was reluctantly moved to a larger site. The old market, however, retained the trade in dairy cattle, pigs, hay, and straw for much of the century. The new market was simply pushed to the city's edge, but remained in the same local area.[27]

The following discussion of the spatial arrangement of Dublin's animal businesses in 1831 and 1901 requires some methodological explanation. The work, undertaken by Dick Tobin and I, involved the correlation of business addresses taken from directories (Pettigrew & Oulton for 1831, Thom's for 1901), with their contemporary coordinates. A range of animal-related businesses were entered into an Excel spreadsheet along with their address at the time and matched with a contemporary location. The process was aided by the fact that Dublin's streetscape has been relatively stable and thus many addresses could be discovered through the use of modern Geographical Information Systems. Addresses which could not be assigned automatically to the contemporary city grid were designated by hand using a list of extant addresses and choosing one as close to the historical address as possible. Finally, those that could not be placed in this way were assigned a location by a combination of historical maps and the city directories' records of cross streets. The addresses were pooled together into grid squares representing 2,000 square feet. This allowed us to generate a map distributing the businesses across Dublin and taking note of local concentrations. The following discussion is based on the comparison of maps made of the businesses in 1831 and 1901. The businesses were assigned into categories for better visualization. 'Meat' consists of butcher's shops, slaughterhouses, and provision curers. 'By-products' contains tallow chandlers, skinners, tanners, furriers, curriers, glue, size and parchment makers. 'Horse' includes places which sold horses (repositories) and stored and hired them (livery stables, cab drivers) but also places that manufactured the carriages, saddlery, and harnesses needed to make use of them. Finally, 'Cattle' includes livestock dealers, salesmasters, drovers, and others associated with the trade in cattle (including sheep and pigs). Unfortunately, dairies are excluded from the data as they were not individually recorded until the 1860s.

Limitations of the data must be acknowledged at the outset. The 1901 directory contains many more entries of businesses than the 1831 directory. For 1831, we mapped 334 businesses, while in 1901 we mapped 672. In

26 Richard Perren, *The Meat Trade in Britain, 1840–1914* (London, 1978), pp. 32–3.

27 Clare, 'The Rise and Demise of the Dublin Cattle Market'.

some cases this may reflect the expansion of a trade, but it is more likely to suggest the increased comprehensiveness of the directory. For example, the 1841 census (the earliest available complete census for Dublin) shows a similar number of people employed as coachmakers as in the 1901 census, yet substantially fewer businesses were listed in the 1831 directory than in 1901.[28] We have assumed, based on the examination of intervening directories, that this lack of representation is proportional across categories of businesses. Further, we have also had to assume that the distribution pattern of businesses has not been dramatically distorted by lower levels of data available for the earlier date. Again, we base this assumption on examination of intervening directories which indicate that localization of particular business categories was marked and slow to change. However, this discussion must be treated as indicative rather than decisive.

## By-products

The clustering of some types of businesses in discrete regions of the city was marked. In 1831, the processing of animal by-products was predominant in the south-western sector where there were also a large number of butchers. Tallow chandlers were by far the most abundant of these types of businesses in both 1831 and 1901. The industries for processing by-products were particularly numerous in the district known as the Liberties. During the eighteenth century, the Liberties had been a centre for silk-weaving, but this trade had almost disappeared by the 1830s. The distinct clustering of similar businesses is highlighted by the directory entries for Back Lane and Watling Street which together accounted for the majority of by-product processing. Back Lane was filled with curriers, tanners, and leather manufacturers. Watling Street was almost entirely comprised of skinners.[29]

The by-products industry in Dublin shows some characteristic clustering of noxious trades with undesirable institutions and poor housing. Hospitals and workhouses were often built on the fringes of the city because their existence alone was seen as a health hazard. As a result, they could have the effect of confirming the undesirability of an area and often joined a cluster of obnoxious proto-industries that processed the animal and human wastes of the city. These included knacker's yards, bone boilers, tallow chandlers, and manure depots. At the western periphery of the south-west district in 1831 was the House of Industry and Foundling Hospital, later converted to the

28 Another possible explanation for this difference is that businesses were less consolidated in the hands of large firms in 1901 than they had been in 1831.

29 *Thom's Irish Almanac and Official Directory* (Dublin, 1901).

South Dublin Workhouse upon the introduction of the Poor Law in 1838 to Ireland. The adjacent streets held some of the highest concentration of noxious animal trades, although the largest number was in fact next to the city's two medieval cathedrals and Dublin Castle. By 1901, however, this latter group of businesses had all but vanished.

There was a decided decline in the processing of by-products within Dublin between 1831 and 1901. In the census of 1841, there were over 500 individuals occupied as skinners, curriers, and tanners. A further 300 were employed as chandlers and soap boilers. Many of these did not appear in directories: the directories list business premises and proprietors, not employees. By 1901, there were only eighty men in Dublin processing skins and hides and fewer than one hundred engaged in other types of by-products manufacture, including tallow chandlery.[30]

The disappearance of the by-products trade was not due to the attentions of sanitary officers, despite the fact that complaints about smells from bone boilers and tallow chandlers were routine. It is possible that the fact that these businesses were already concentrated in particular areas, and particularly poor areas, meant that they were generally ignored. The corporation minute books indicate that they were only infrequently viewed as priority nuisances. In 1856, for example, the sanitary officer inspected the Watling Street area and found that although there was a 'heavy smell' in many of the premises, it was not perceptible from the street.[31] The same officer declined to prosecute bone boilers, despite a medical man's testimony that they were 'prejudicial to health', or to suggest the removal of soap and candle makers beyond the borough despite appeals from urban residents. In each case, the officer found the industries to be both necessary and without ill effects.[32] The exception was knacker's yards—the places for disposing of horses—of which there were only two in the city. In 1869, the corporation refused to grant a permit for a further yard on the grounds that there was no demand for it and that for the sake of health they should be kept to a minimum.[33]

The by-products trade appears to have been decimated by a declining demand and other economic forces rather than sanitary reform. The flourishing Irish cattle trade exported its product alive, requiring no processing of offal in the city. Proponents of developing an Irish trade in dead meat rather than livestock at the close of the century found that an absence of processing for

30 Ireland, Census Office, *Report of the Commissioners Appointed to Take the Census of Ireland, for the year 1841*, p. 22, table VI; *Census of Ireland, 1901*, Part I, vol. 1, *Province of Leinster*, p. 25, table XX.

31 Dublin City Council, Public Health Committee minute books, vol. 2, entry for 8 Aug. 1856, Dublin City Library and Archive (hereafter DCC PHC mins).

32 DCC PHC mins, vol. 5, entries for 19 Sept. 1856 and 6 Feb. 1867.

33 DCC PHC mins, vol. 6, entry for 7 May 1869.

offal was an obstacle.[34] What offal there was from the local meat trade does not appear to have stimulated substantial business interest. Instead, the bones were 'left to putrefy' and eventually converted to cheap manure by bone boilers.[35] The blood was collected and sent to Manchester and Glasgow for processing.[36] Other bits of edible offal were evidently given to slaughterhouse employees for their own consumption. Only the fat, or tallow, continued to be routinely processed in the city in a widely distributed network of chandlers and soap makers.

## Meat

The processing and sale of meat exhibited a different, though related, spatial pattern from that of by-products. Of all urban animal industries, the slaughterhouse has been most frequently subjected to geographical scrutiny in the context of the nineteenth-century city. At the start of the century, urban slaughterhouses clustered with areas of low-value housing, but like such housing they were also remarkably widely distributed in a number of cities.[37] Efforts to centralize slaughtering in supervised public abattoirs and to remove these from the city centre formed part of the plans of city governors in Europe, Britain, and America. Urban growth meant that even sites originally removed from population were often swallowed up by housing in new phases of growth. Yet, by the close of the nineteenth century, most cities in the United Kingdom and many in Europe and America had consolidated slaughtering to a minimal number of large abattoirs. These abattoirs allowed not only for the removal of unpleasant sights, sounds, and smells from city neighbourhoods but also for greater supervision by municipal authorities in an effort to control the sale of diseased meat. The full industrialization of meat production in many places, including Dublin, was held back by slow adoption of refrigeration technology.

34 D. Tallerman, *'Railway Abattoirs' and Other Papers Relating to Meat Distribution* (Dublin, 1891), p. 90.

35 Ibid.

36 House of Commons Parliamentary Papers, *Report of the Royal Commissioners Appointed to Inquire into the Sewerage and Drainage of the City of Dublin*, 1880, minutes of evidence, evidence of Charles Cameron, p. 51.

37 Prunty, *Dublin Slums*, 91; I. MacLachlan, 'A Bloody Offal Nuisance: The Persistence of Private Slaughter-Houses in Nineteenth-Century London', *Urban History*, xxxiv (2007), pp. 227–54; C. Otter, 'Cleansing and Clarifying: Technology and Perception in Nineteenth-Century London', *Journal of British Studies*, xliii (2004), pp. 40–64; D. Brantz, 'Animal Bodies, Human Health and the Reform of Slaughterhouses in Nineteenth-Century Berlin', *Food & History*, iii (2005), pp. 193–215; P. Young Lee, 'Hide, Seek, Slaughter, Meat: The Slaughterhouse as Site', *Food & History*, iii (2005), pp. 241–90.

The distribution of Dublin's butchers and slaughterhouses in 1831 and 1901 changed slightly, but did not substantially shift towards the urban periphery. The erection of a municipal abattoir in 1882 had little immediate impact. Through licensing and regulation, the corporation managed to force the closure of almost half of the city's private slaughterhouses between about 1878 and 1901. At the same time, the numbers of people employed as butchers and victuallers in the city actually rose steadily up to over 800 individuals by 1901.[38] We have to assume that many of these persons were simply meat sellers. The slaughterhouses that stayed were generally allowed to remain in clusters and therefore their pattern within the city was remarkably consistent over time. A tract encompassing Thomas Street/James's Street, and the streets immediately to its south, running parallel to the river in the south-western district, was an important area for the production of meat in both 1831 and 1901. There was a distinct and persistent repulsion of meat businesses from the salubrious south-east quarter, with the exception of lower-value lands north and east of Trinity College Dublin and near to the railway lines.

Other factors probably influenced the stability of particular trades in specific urban locations. The Dublin Corporation sought actively to rid the city of what it considered the significant nuisance of the slaughterhouse located at the back of a butcher's shop in a residential neighbourhood. These types of businesses were resistant to movement for a wide range of reasons, but one was simply related to the buildings themselves. A slaughterhouse and butcher's shop, already outfitted for trade, represented a considerable outlay by the owner and likewise an attractive purchase for a new owner. A butcher needed a yard of a certain size to keep animals in, a distinct space for slaughter, access to water to clean the facility, a butcher's shop, and perhaps a house. For example, an advertisement offering a butcher's premises for sale in 1862 described a large house with over six rooms, a yard, sheds for cattle, blocks and choppers, as well as a constant supply of water.[39] A premises for slaughtering pigs and curing bacon for sale in 1866 had similar special facilities, including a large walled yard, a shed, and a gate entrance separate from the dwelling house.[40]

A small movement of meat selling and processing to the north of the city was spurred on by two commercial projects with sanitary aims. In 1877, the South City Markets Company bought several city blocks west and south of Trinity College which contained a number of butchers and slaughterhouses. They razed the buildings and permitted only a few butchers in the newly

38 *Report of the Commissioners Appointed to Take the Census of Ireland, for the Year 1841*, p. 22, table VI; *Census of Ireland, 1901*, Part I, vol. 1, *Province of Leinster*, p. 25, table XX.

39 'To salesmasters, victuallers, poulterers, provision dealers, &c, &c. For sale by auction', *Freeman's Journal*, 20 Aug. 1862.

40 'Auction of pig premises', *Freeman's Journal*, 21 Feb. 1866.

erected market, thus diminishing a small cluster of inner city meat businesses.[41] In 1896, the Bull Alley area was razed. Formerly a large centre for meat processing and selling, and also perceived as a den of vice, Bull Alley was replaced with homes erected by the Artizan Dwellings Company. Nonetheless, these acts neither diminished nor consolidated butchery in the city. Instead, established areas such as Moore Street on the north side and Camden Street on the south side appear to have expanded. The production and sale of meat remained widespread throughout the city and was certainly not excluded from densely populated districts. In 1901, the butchers were more widely distributed and found at the edges of the metropolitan district, but not at the expense of city centre butchers. The new cattle market at the north-western corner of the city encouraged the establishment of local butchers, but, reflecting the decline discussed above, few places for processing by-products.

## Horses

Dublin was apparently dubbed 'the most car-driving, tea-drinking city in the universe' sometime in the 1830s. Guidebooks and Dubliners alike embraced this description and continued to proclaim its truth even after the introduction of railways and electric trams might have been imagined to reduce the demand for horses. In the 1830s, a visitor made the unlikely claim that 'all the coachmakers were in such full employment, that no contract could be obtained for building coaches on the Dublin and Kingston railroad'.[42] In fact, new forms of transport tended to increase the need for horses, and it was well into the twentieth century before horse-drawn transport began a marked decline. The horse's special status as a 'clean' animal made it an acceptable presence even in the most salubrious neighbourhoods. Horse-related businesses were at least as widely distributed as those related to meat and they were found even in areas where no other animal businesses penetrated.

The flight of wealthy Dubliners to the suburbs and the declining value of even the most grand Georgian squares made space for horse businesses. Livery stables and veterinary surgeries moved into the generous stables and yards of formerly grand homes. In 1901, there was at least one such business on each side of St Stephen's Green and on two sides of Merrion Square. The squares faced onto parks and each row of houses had a stable lane behind it. The residents of the squares were less and less likely to own horses themselves; instead, former homes provided commercial space on a scale suitable to keeping horses. The lanes and alleys of the commercial districts (south of

41 DCC PHC mins, vol. 14, entry for 21 Dec. 1877.
42 Henry Inglis, *Ireland in 1834* (3rd edn., London, 1835), p. 9.

Trinity College and near to Sackville Street) also provided space for city shoppers to stable horses or to attend to shoeing.

Horse businesses were widely scattered in both 1831 and 1901, but they clustered near to roadways leading to the city's edges. Many of these same roadways became the basis for horse-drawn tramlines. For example, one detectable concentration in 1831 was on either side of the line I have designated as dividing the eastern and western quarters of the city. This was one route to the developing suburbs and also to the Portobello Barracks outside the Circular Road. The pattern of horse businesses lined along the roads connecting the city to the countryside was even more marked in 1901, when businesses proliferated along routes to the Phoenix Park (in the north-west) and also along roads to the city's north-eastern and south-eastern suburbs. Such a pattern is perhaps not surprising for an animal used for transportation.

Clusters of horse businesses can also indicate areas of high demand for transportation or concentrations of like trades. For example, in both 1831 and 1901 there were large numbers of cab drivers and car owners clustered in the area surrounding the King's Inns (the Irish law school), where they presumably found employment bringing lawyers between offices, courts, classrooms, and home. The increased demand for horses as adjuncts to railway transport is also seen in their proximity to several railway termini. The growth in railways actually expanded the role for horses in the city to include bringing passengers and goods between stations and into town as well as working in the railway yards.[43] Horse businesses were also associated with the cattle market area in the north-west (both the new market and old Smithfield), where they presumably served not only Dubliners, but farmers while they were in the city to sell stock and produce. Finally, an east–west line of businesses almost entirely devoted to horses was discernible in 1831, north of the river. The line was pronounced in 1901 when Great Britain Street (now Parnell Street) and a large portion of the north-eastern city were dotted with carriage makers, saddlers, harness makers, livery stables, veterinary surgeons and farriers.

## Cattle

Another trend confirmed by the maps is the growing impact of the Dublin cattle market. In 1831, the market was located at Smithfield, just a short distance north of the Liffey in the north-western district. A small cluster of livestock-related businesses, principally salesmasters, surrounded the market. By 1863, the market had been moved to the very north-western edge. As

43 C. McShane and J. Tarr, *The Horse in the City: Living Machines in the Nineteenth Century* (Baltimore, MD, 2007).

the market boomed, so did the cattle dealers, salesmasters, and drovers, the majority of whom now settled in adjacent streets. Many retained business premises at Old Smithfield, but newcomers lined the streets leading to the market. Nevertheless, the overall influence of the city's livestock trade is reflected in the scattering of cattlemen throughout the city with the exception of the south-east sector.

The development of Dublin's north-western district, particularly the part comprising Arran Quay Ward, appears in association with the expansion of a wide variety of animal businesses. In 1831, the north-west appeared as a relatively empty sector (although the map excludes the Ormond Market as previously mentioned). By 1901, it was clearly a booming district with industries catering to the cattle traders. This included not only those who dealt directly in the market and in animals, but also hoteliers and publicans who housed and fed the market's human contingent, butchers making use of the market's produce, and horse businesses presumably profiting from the increased traffic to the area. The exception to the north-west's dominance in cattle is North Wall Quay, in the far north-east of the city. It was here that cattle were exported to England, and thus another cluster of cattle-related businesses was well-established there by 1901. These included steam ship companies dealing predominantly in livestock, lairs and yards used to contain them, and the offices of cattle dealers who sold to England.

## Experiencing the Animal City

Through a mapping of animal businesses we can begin to get a sense of the city's environmental landscapes. A further question to ask is how Dubliners experienced these landscapes and whether or not they reflected upon them. Certainly reformers were concerned with the presence of polluting businesses and the Dublin Corporation reports contain many complaints of pollution from ordinary citizens. While the landscape of animal businesses did not change dramatically over the period surveyed, what seems to have changed are attitudes towards them. I detect a hardening attitude towards the presence of particular animals in the city over the nineteenth century. This is evidenced in the proliferation of the city's public health activities, in the outpouring of public health commentary by medical men, and also by the reflections of urban residents themselves. Historians of the senses have suggested that what seems a biological process to us, such as smell, may be culturally constructed. What smells 'bad' to humans varies across cultures and also through time.[44]

44 For an example in the Irish literature, see J. Nugent, 'The Human Snout: Pigs, Priests, and Peasants in the Parlour', *Senses and Society*, iv (2009), pp. 283–302.

This appears to have been the case across the nineteenth century in Dublin. Further investigation is required, but the comparison of two perambulating Dubliners (one real, one fictional) may open up the discussion.

Two examples that illustrate the point are the 1841 diary of James Christopher Kenny (a wealthy Catholic student) and the twentieth-century perambulations of Leopold Bloom (the fictional protagonist of James Joyce's *Ulysses*). Both gentlemen ranged far and wide across what was a compact city. Until the inclusion of the suburban townships in the Dublin municipal district (from 1896), the area that was under control of the Dublin Corporation could be traversed on leisurely horseback in about an hour. The expectation that a city sanitary official could visit, on foot, over one hundred slaughterhouses scattered throughout the borough every week suggests that it was (and is) equally walkable.[45] Many Dubliners, even some residing in the more distant suburbs, walked to and from work throughout the century.[46]

Kenny mostly experienced Dublin on horseback, a fact that was shaped by his social and cultural status. Kenny was a Catholic student of law at Trinity College in his mid-twenties and he was privileged enough to have personal use of a horse. His family rented a house in Blackrock for £40 per annum where they kept a stable of horses and their own carriage.[47] He often recorded his journeys in and through the city on foot and horseback. On one particular journey Kenny travelled from Blackrock on the south-eastern fringe of the city towards the rural village of Tallaght on the western fringe. He passed through the centre of town and then headed west over the Grand Canal and beyond the municipal boundary. Kenny's journey crossed one of the city's main socio-economic gradients, that of east to west. He moved through a variety of urban habitats from his own neighbourhood of wide streets and spacious houses, through the densely populated Coombe full of low-value housing and small industries, and into the proto-industrial outskirts along Cork Street.[48] Matching Kenny's route with Ordnance Survey maps and directories of businesses underscores the contrasting status of transport and food animals. Horse-related businesses were relatively widely dispersed throughout the city, but other animal businesses were almost exclusive to slum areas. As Kenny approached Merrion Square and St Stephen's Green from Blackrock, he passed livery stables, Robert Bohy's 'Royal Horse Bazaar' and other horse dealers, a veterinary surgery, and carriage makers. These types of businesses

45 DCC PHC mins, vol. 6, 17 May 1867.

46 Diary of William Woodlock for 1865, NLI MS 3779; M. E. Daly, 'The Growth of Victorian Dublin', in M. E. Daly, M. Hearn, and P. Pearson, *Dublin's Victorian Houses* (Dublin, 1999), p. 19.

47 Diary of James Christopher Kenny, MS 10800, Trinity College Dublin; see entries for 25 Aug. and 12 Nov. 1840.

48 Kenny diary, 24 Sept. 1840.

continued, while decreasing in numbers and in value, throughout the Coombe and Cork Street areas. They were joined by slaughterhouses, bacon curers, dairy yards, victuallers, and tanneries. The open and green spaces of wealthy, residential Dublin in the east appeared in stark contrast to the crowding and dirt of the poor districts of the inner city. Kenny passed these changes without remark. In fact, aside from his own horse and pets, the animals of the city appear to have been largely invisible to Kenny. The stench of manure depots, the lowing of cows in urban dairy yards, the squealing of pigs being castrated or slaughtered, and the scuttling of rats were likely to have been familiar smells, sounds, and sights that did not merit comment.

More than sixty years after Kenny, James Joyce's Leopold Bloom made his many pilgrimages across and through Dublin. The city remained compact and its social divisions similar. Bloom covered vast tracts of Dublin in a single day, mostly on foot but occasionally making use of other forms of transport. A shared carriage journey to a funeral took him from the city's south-eastern edge to its northern one, thus traversing the city's other socio-economic division of north and south. Along the way he passed from the salubrious Sandymount Green into the relatively desolate South Dock area, where he noted passing the home for stray dogs and cats. The carriage then crossed the river and passed along the shopping thoroughfare of Sackville Street. The carriage paused in Berkeley Street, its path blocked by cattle being driven towards the river for export, and then passed over the canal. Just as Kenny's journey brought him closer to the countryside where his lover now lived, Bloom's journey provided a visible connection between Dublin and the countryside in the form of cattle and the canal. Bloom thought of the canal, 'Dropping down, lock by lock to Dublin', as a tangible line connecting him to his daughter in the rural town of Athlone.[49] Although the timing of the journey is uncertain, the conversation suggests that Bloom's ride was not particularly long. Like Kenny, he had quickly traversed several city districts associated with different social, economic, and cultural spheres.

While Kenny and Bloom's journeys show a city whose social geography and size had changed little, their remarks about their surroundings tentatively suggest a shift in ideas about the place of animals in the city. While Kenny ignored the city's non-human residents, Bloom was constantly aware of them. Animals had not diminished substantially as an element of daily life in the intervening years, but human residents may have questioned their presence more. Upon seeing the cattle, Bloom remarked, 'I can't make out why the corporation doesn't run a tramline from the parkgate to the quays ... All those animals could be taken in trucks down to the boats'. He further suggested replacing the traditional 'hearse and carriage' with special funeral

49 James Joyce, *Ulysses* ([1922] London, 1998).

trams which would be 'more decent than galloping two abreast'.[50] Bloom had been employed by the cattle market and the facts of slaughter and meat production recur throughout the novel. Indeed, the exploitation of animals in the city is a prominent enough theme in Joyce's novel to merit a 'vegetarian reading' of it.[51]

## Conclusion: Animals, Environment, and Modernization

Nineteenth-century cities were the focus of much reforming and modernizing zeal and Dublin was no exception. The improvement of the urban environment for its human residents was a common refrain. Yet the imposition of certain visions of improvement was impeded by a variety of factors. One was the environment and landscape of the city itself. In the case of Dublin its rivers contributed to a socio-economic pattern that persisted despite new regulations and development. Animal businesses were also slow to remove to the urban periphery, mirroring many other cities.[52] We might predict that between 1831 and 1901, Dublin's animal businesses would tend to diminish in number and those remaining would move further to the periphery. Although some movement did indeed take place, the city in 1901 was as dominated by animal industries as it had been in 1831, if not more so. This chapter has highlighted a number of reasons for resistance to change including the accessibility of resources for 'dirty' businesses (e.g., water supply) and capital investment in business premises. Nonetheless, the attitudes of many in government and of the middle classes more broadly changed over the century.

Considerable research remains to be done on sanitation in the city, but we must consider the complex interplay between the public health campaign and other factors. For example, city dwellers and businessmen often had economic reasons for failing to comply with sanitary regulations. The Dublin Corporation succeeded in establishing a public abattoir on the outskirts of the city in 1882, but it took another several decades before private slaughterhouses had been mostly closed down and butchers forced to use it. Butchers claimed that in-house slaughtering was both a business necessity and more reliable than the abattoir. The public health act of 1878 set a minimum distance of fifty feet between human habitations and animal pens. Yet urban sanitary inspectors continued to remove pigs from dwelling houses and yards well into

50 Ibid., p. 96.

51 Peter Adkins, 'The Eyes of That Cow: Eating Animals and Theorizing Vegetarianism in James Joyce's *Ulysses*', *Humanities*, vi (2017), pp. 1–15.

52 MacLachlan, 'Bloody Offal Nuisance'; Philo, 'Animals, Geography and the City'; Otter, 'The Vital City'.

the twentieth century because keeping a pig continued to be a useful means of generating cash income among the poor.[53] The result of these tensions was a very patchy 'modernization' of the variety favoured by reformers. However, regulations existed by the close of the nineteenth-century that would enable the slow removal of animals and animal businesses from the city. As one of Dublin's last pig dealers reported towards the end of the twentieth century, 'It's a known fact that the Corporation want piggeries out of Dublin. That's been going on since me father's time. It's a slow process'.[54]

The urban environment of nineteenth-century Dublin was not just a human one. As Kevin Kearns wrote, 'in the first half of the century Dublin streets could be so clogged with animals that it looked like fair day. Animal life was a normal part of city life'.[55]

Kearns was speaking of the twentieth century; the city's reliance on non-human animals left its mark long after the nineteenth century ended. A more comprehensive environmental history of the city would also consider the impact of changes such as the introduction of flush toilets and the main drainage scheme, alterations of the quays and port, and the building of roads and railway lines. As I believe this brief introduction with a focus on animals has suggested, urban change results from the negotiation between human and non-human. As Frederick L. Brown has argued for Seattle, 'The city is more than human'.[56]

53 See *Report of the Committee Appointed by the Local Government Board for Ireland to Inquire into the Public Health of the City of Dublin* (Dublin, 1900), 13; *Minutes of Evidence Taken before the Committee Appointed to Inquire into the Public Health of the City of Dublin; with Appendices* (Dublin, 1900), pp. 165–6.

54 Quoted in K. C. Kearns, *Dublin Street Life and Lore* (Dun Laoghaire, 1991), p. 191.

55 Ibid., p. 57.

56 Brown, *The City Is More than Human.*

# 8

# Mainstream or Tributary? The Question of 'Hibernian' Fishes in William Thompson's *The Natural History of Ireland* (1849–56)

*Mary Orr*

The main question and arguments of this essay are encapsulated in its title. It examines how geographical ('Hibernian') and scientific ('mainstream', 'tributary') terminologies and systems of classification need to be historicized in a specific period (mid-nineteenth century) on two important accounts. First, the frames of reference for such terminologies and classifications may be different from today. Second, these frames of reference also shaped wider intercultural interactions and exchange. In using seemingly ahistorical river terms and metaphors and classical Latin/'poetic' geographies, the title thus draws attention to the need for more careful contextualization and questioning of assumptions about 'Ireland'. For example, the development of geology, hydrogeology, and ichthyology as major disciplines and new sub-disciplines in nineteenth-century scientific endeavour brought revised scientific terminologies for river systems. An indicative contemporary definition below thus frames this essay, to locate its wider ramifications. In offering the first evaluation of the contributions of William Thompson's *The Natural History of Ireland* (1849–56) in the history of nineteenth-century ichthyology, this study therefore also re-evaluates the status of 'Ireland' in the history of nineteenth-century natural history. By arguing that both Thompson (1805–52) and (nineteenth-century) Ireland merit a more centrally contributory rather than auxiliary positioning in the history of scientific endeavour, the essay challenges the use by historians of allegedly a-temporal river metaphors: they are never culturally or historically neutral. In showing how Thompson's work occupies a more central place on the nineteenth-century scientific map as an important, comparative case study, the essay can then conclude with the 'modest proposal'[1] that overtly territorialized and overly terrestrial conceptions of natural history endeavour lose sight of more significantly fluid and inter-connective scientific and cultural understandings of things.

According to Thompson's contemporary, the differently overlooked popularizer of science, Rosina M. Zornlin:

> The main or principal stream is designated the *recipient* stream, because it receives the other streams. ... Rivers which flow into the recipient, are termed *affluent streams*, because they flow towards, and directly into, the recipient stream. ... In some instances two rivers unite their streams, and the names of both are lost in a new appellation; thus forming what are termed *confluent* streams.[2]

Since rivers and their inhabitants follow changing watercourses irrespective of the primary, national, or other terrestrial status of these, fish species and their scientific understanding are thus also subject to shifting and overlapping temporal, as well as geographical and political, definitions and classifications. The comparative contents and conceptual status of the fifth and final appendix of William Thompsons's four-volume *Natural History of Ireland*, entitled 'Fishes of Lough Neagh and Lake Geneva', to which we will return, neatly illustrates the point. The same 'Hibernian/Irish' fishes inhabit both these locations making them equally 'Swiss'. In a greater body or structure, however, an appendix is always a subsidiary adjunct: in a publication, it is 'an addition subjoined to a document ... having some contributory value, but not essential to completeness' (*OED*). In respect to the impacts of Thompson's work, the cartographical specificity of his title has thus clearly created similar 'appendix-effects'. At the time of its publication, the 'single' nation jurisdiction and designation, 'Great Britain and Ireland', made Thompson's field(s) of endeavour geographically, politically, and grammatically appositional, with subsequent reception seen as derivative of allegedly more important contributions to knowledge. In the

1 The greater significance of Jonathan Swift's much-cited title for the history of natural history will be addressed in the conclusions.

2 R. M. Zornlin, *The World of Waters; or Recreations in Hydrology* (London, 1843), p. 228 (emphasis in the original). Bernard Lightman has collected Zornlin among important but forgotten popularizers in need of rediscovery. Her uncertain dates in his listing of her in his 'Marketing Knowledge for the General Reader: Victorian Popularizers of Science', *Endeavour*, xxiv (2000), pp. 100–106 (p. 102) perhaps explain why her name is missing from the *Oxford Dictionary of National Biography*.

nineteenth-century public mind, 'Ireland' as the smaller adjacent land mass of the British Isles remained at best the lesser, detachable knowledge domain, an adjunct or periphery of the greater Great Britain of England, Scotland, and Wales. At worst, Ireland was ever the backward backwater, defying civilizing projects, including the scientific. Even when Ireland became an independent nation state and member of the EU almost a century later, there was little concomitant shift in the twentieth-century *critical* mind regarding its independent, or interdependent, knowledge value.

The potential importance of Thompson's work has thus only resurfaced in the last twenty years in reappraisals dedicated specifically to recovering Ireland's heritage as an important, because separate, geographical and cultural entity. A milestone was *Nature in Ireland: A Scientific and Cultural History* in 1997.[3] Although no chapter makes Thompson its object, David Cabot calls his *Natural History of Ireland* a 'zoological landmark ... still used today as a source of reliable information'.[4] Christopher Moriarty's study 'Fish and Fisheries' analyses the importance of the latter in the West of Ireland, but does not develop the point that 'While William Andrews studied fisheries, William Thompson was studying the fishes'.[5] Of the small number of scientific articles specifically addressing aspects of Thompson's work, none exceeds six pages, and most appear in the *Irish Naturalist's Journal*.[6] Their slightness only adds to perceptions of his rather minor, and localized, historical and scientific significance. The appeal of Thompson's *Natural History*, like the appendix, thus remains circumscribed by niche audiences with research interests in 'Ireland' understood as an outpost geography, a set of sub-regions before/after partition, or as constituting sub-disciplinary knowledge interest—for example, 'Irish ornithology'. Ireland's longer or larger cultural and scientific heritages thus remain landlocked as local, antiquarian, folkloric, nationalistic, or diasporic concerns, rather than as nationally or internationally significant. It is a known fact, for example, that Ireland's flora and fauna today, as in the nineteenth century, are fewer in number than on the 'greater' British side of the Irish Sea.[7]

3 J. W. Foster and H. C. Chesney, eds., *Nature in Ireland: A Scientific and Cultural History* (Dublin, 1997).

4 D. Cabot, 'Essential Texts in Irish Natural History', ibid., pp. 472–96 (p. 491).

5 Ibid., pp. 283–98 (p. 291). See also John Roney's chapter, ibid.

6 See M. W. Rae, 'The Wm Thompson Collection of British Marine Algae', *Irish Naturalist's Journal*, v (1934), pp. 81–3; O. Morton, 'Three Algal Collections in the Ulster Museum Herbarium', *Irish Naturalist's Journal*, xx (1980), pp. 33–7; P. N. W. Jackson, 'William Thompson (1805–1852): Zoologist and Biogeographer', *Irish Naturalist's Journal*, xxx (2009), pp. 119–22.

7 See N. Maclean, ed., *Silent Summer: The State of Wildlife in Britain and Ireland* (Cambridge, 2010). For fishes, see P. S. Maitland and J. F. Craig, 'Freshwater Fishes: A Declining Resource', ibid., pp. 383–400.

The direct result is that William Thompson's substantial contributions to nineteenth-century natural history and, importantly, to its history and legacies, have barely figured in modern British, that is mainstream, Anglophone history of natural science. David Elliston Allen's eminently comprehensive *Books and Nature* (2010) is illustrative of such bias, and of an unconsciously prejudicial view of 'Irish' contributions. In a chapter highlighting the importance of nineteenth-century publishers of natural history works Allen only mentions Thompson tangentially through his publisher Reeve, despite acknowledging Thompson's reputation as the 'leading Belfast naturalist of the day':

> Reeve had had no better luck in the meantime with another equally worthy undertaking. This was the comprehensive account of the Irish fauna, extending to many octavo volumes, that the leading Belfast naturalist of the day, William Thompson, was to produce. All too soon, however, after only three of the promised volumes had come out, and those covering just the birds (and admirably thoroughly), the author died unexpectedly, in 1852, when still not yet fifty. **Short-changed** already **by a misleading title**—for where was the botany, let alone the geology, in a work **claimed as** comprising the *Natural History of Ireland?*—the subscribers received **a further, almost insultingly short measure** as a result, and that after a five-year wait: **only a single volume more**, into which **the whole of the rest of the fauna was cursorily crammed**, the best that two of Thompson's friends had been able to do with **merely fragmentary notes of his** to go on.[8]

The emphasis not only illuminates Allen's extraordinary castigation of Thompson for deceptions of every kind, including his failure to complete and deliver his *Natural History of Ireland* from beyond the grave. It also epitomizes how critical dismissal of Thompson and of Irish natural history endeavour can elide in a single stroke. The alleged incompleteness of Thompson's work in the 'cursorily crammed' fourth volume excuses examination of their contents (let alone final appendices) as also indicative of the quality of the three finished volumes. Indeed, by over-writing Thompson's achievements and title for 'completeness', Robert Lloyd Praeger's definitive *Natural History of Ireland* of 1950 acknowledged only to dismiss its precursor in the lineage of the history of Irish natural history.[9] Praeger was the grandson of Robert Lloyd Patterson,

8 D. Elliston Allen, *Books and Nature* (London, 2010), p. 215 (emphasis added). It lies outside this chapter to map the reactions of Thompson's subscribers and readers of his work in Britain during his lifetime.

9 This is a textbook case of Bloomian 'anxiety of influence', but for natural history rather than for poetry. See H. Bloom, *The Anxiety of Influence: A Theory of Poetry* (Oxford, 1973).

Thompson's close friend and main editor, who faithfully brought volume 4 of the original *Natural History of Ireland* to publication.

Zornlin's definitions above of 'recipient' and 'affluent' river sources, however, understand parts (tributaries) and wholes (mainstreams) very differently from their assumed hierarchies of importance. By following her definitional cues, this essay now takes an altogether different course to the work of earlier critics, including those overtly promoting the rediscovery of Thompson. First, its focus on fishes will demonstrate how the final and complete appendix of *Natural History of Ireland* is in fact a summa of Thompson's foremost knowledge of his subject and its treatment, the better to promote Ireland's wider-reaching, transnational significance. How then did he contribute directly and fully to the very new discipline of nineteenth-century ichthyology, and hence to the ichthyology of Ireland as constituent of it? Second, how does study of Thompson's work on fishes remap twenty-first-century terms of engagement with comparative, that is *'confluent'*, natural science endeavour as a more interestingly complex, and intercultural, phenomenon?

### *'Fishes of Lough Neagh and Lake Geneva'*

In the viewfinder of Zornlin's 'fluvial' terminology, Thompson's final 'appendix' provides a subject-defining supplement that affluently informs its larger wholes, whether *Natural History of Ireland*'s 200-page section on fishes, its multi-volume survey, or its geography within a greater Great Britain/Europe. The status and stature of Thompson's contribution on the fishes could not in fact be clearer in terms of completeness and quality of substance, thanks to the unusual intervention and endorsements of his Editor in a note qualifying the appendix title:

> [It seems to have been Mr. Thompson's intention to have investigated the Natural History of Lough Neagh and the Lake of Geneva, both positively and comparatively; but that part of his MSS. **which treats of the fishes, is the only portion which has been left in a state sufficiently far advanced to warrant publication.**—Ed.][10]

The all-important semi-colon here brings together the encyclopaedic reach of both appendix and sectional coverage—number of species, scientific breadth and depth of information gathered for each—as an inter-reflecting whole. The preceding 200-page survey of the fishes draws together Thompson's major

10 William Thompson, *The Natural History of Ireland* (4 vols., London, 1849–56), iv. p. 494 (emphasis added).

scientific endeavours investigating marine, estuarine, and freshwater habitats by collating and expanding his many expert publications over some twenty years in the newcomer science of ichthyology. The final appendix is thus illustrative of Thompson's established authority within this branch of natural history just as its pithy scope, structure, comparative methodology, and *international* specificity encapsulate what his larger *Natural History of Ireland* proposes. Thompson was acutely aware of the ingrained, and prejudicial, devaluation of Irish subjects—whether people, knowledge, culture, or natural history—within British Gentlemanly Science.[11] He therefore set his project proactively within the foremost agendas of nineteenth-century European continental discovery and debate in natural scientific inquiry, to which Gentleman Science also aspired. Lough Neagh comes first in the title of the final appendix strategically. As the largest inland freshwater lake in Ireland *and* the British Isles it has in apposition and for direct comparison the largest inland freshwater lake in Switzerland and central Europe. These two freshwater bodies formed as part of trans-European river systems that developed at the retreat of the last Ice Age. *Contra* Allen, above, geology is therefore everywhere present in Thompson's *Natural History of Ireland* as the underpinning explanation for Ireland's many distinct and distinctive terrestrial, coastal, and fluvial habitats and their many inhabitants. Although fossil species such as the Irish Elk are included in his appraisal of Ireland's former and current fauna in volume 4, Thompson's primary interest is to map and understand for the first time the richness and diversity of the many different life and living forms that ultimately date back to the geomorphological separation of 'Hibernia' from continental Europe. For Thompson, the many complex niche habitats in which Ireland is particularly rich—its local 'eco-systems', in current parlance—denote at once sub-regional, regional, and supra-regional phenomena depending on the optics of vision for the species in question.

By correlating the two major inland lakes that formed in Europe at the same period of retreating ice sheets, Thompson can thus draw up three species lists in this appendix, which I have tabulated the better to illustrate his point (see Table 8.1). At the supra-regional level are 'Continental European' freshwater fishes common to both lakes, although Thompson overtly avoids all use of nationalizing geographical designations. The species found only in one or other of these major inland water bodies then indicate differences in their (local) habitats due to longitude, but also post-glacial *bathymetrical* formation. In AD 563, Lake Geneva suffered a major tsunami (caused by a major rock fall where the Rhône enters its east end) that altered its depth and soil deposits. Lough Neagh since its formation has remained unusually shallow for such a

11 See J. Morrell and A. Thackray, *Gentlemen of Science: The British Association for the Advancement of Science* (Oxford, 1981).

large water body.[12] Thompson's intentions are thus transparent: 'a comparison of the two lakes is very interesting, not only as illustrative of geographical distribution, but of the comparative **value** of their finny inhabitants'.[13] This value is scientific and socio-economic (cultural). By using the Latin names (and Linnaean binomial classification system), Thompson firmly situates his work on fishes within international rather than national or vernacular communities of natural history knowledge, although the latter are not ignored.[14]

Readers from or outside Ireland are thus assumed to share the author's scientific fluencies (including knowledge of geology) to benefit from Thompson's gloss on this 'data', extrapolated from multiply informed scientific evidence, including his own personal observations. To quote Thompson's final appendix:

> Here we find the general result that might be anticipated from the geographical position of the two lakes, the more northern being the richer in species of the genus *Salmo*; the more southern in species of the *Cyprinidæ*; for it is well known that **in continental Europe** the *Salmonidæ* increase in number northward, and the *Cyprinidæ* southward.
>
> The **value** of the fishes of the northern lake is vastly greater than that of the southern. The most striking difference in connexion with the species which are common to the two lakes is, that *eels*, which are rare at Geneva, are abundant at Lough Neagh.[15]

The factual baldness of Thompson's lists in Latin and measured style in this appendix reflect his emphases throughout *Natural History of Ireland* on exact and exacting scientific information. Moreover, in promoting and extending

12 See Zornlin's 'picturesque' account of Lough Neagh, further demonstrating prejudicially negative representation of its significance: 'The principal lakes of Ireland are *Lough* or *Lake Neagh*, and the beautiful *Lakes of Killarney*. Lough Neagh is the largest lake in the British Isles, its length being about eighteen miles, and its breadth about eleven or twelve. *This lake is wholly deficient in picturesque beauty, its shores being low and flat, and in some parts marshy, and subject to occasional inundations. Differing thus in character from the beautiful Scotch lakes*, we may expect that it will also differ in the depth of its basin; and accordingly we find, that in no part does the depth of this lake exceed 102 feet'. *The World of Waters*, p. 288 (emphasis added).

13 Thompson, *Natural History of Ireland*, iv. p. 494 (emphasis added).

14 Thompson includes the accepted nomenclatures for Irish fish species used by local experts such as fishermen and fishmongers as part of the scientific completeness of his survey. See, for example, *Natural History of Ireland*, iv. p. 201: 'The Brill or Britt†' ['†Called 'Britt' in Belfast market'], and iv. p. 203, where the Irish name is the only Anglophone term available, 'The Whiff*' ['* Called "she sole" in Belfast market; "ox sole," and also "white sole," in Dublin market'].

15 Ibid., iv. p. 495 (bold emphasis added).

knowledge of Ireland's particular rich fish biospheres to facilitate their wider comparative understanding—exemplified here in Lough Neagh and Lake Geneva—Thompson immediately redraws nationally bounded designations for such species, and hence their assumed hierarchies of value. Lough Neagh's fishes no longer have secondary positioning within 'Great Britain and Ireland' because they validate geologically similar species in now separate 'Hibernian-British' and 'Swiss' continental habitats. Similarly, Thompson's recognition of 'common' and 'distinctive' species in Ireland is thus not only for themselves, or to valorize the latter. Rather their scientific interest as noted here is the northerly predominance of the salmonid fishes, yet their unusual variations in comparable habitats. We will return below to the 'distinctive' forms of *Coregonus* marked in bold in Table 8.1. The 'common' are no less vital markers for species ranges across time, regionally and transregionally, to understand the similarities, differences, and changes in expected patterns.

**Table 8.1 'Fishes of Lough Neagh and Lake Geneva', from William Thompson's *The Natural History of Ireland* (1849–56), iv. 494–5**

<table>
<tr><th colspan="2">Lough Neagh and Lake Geneva</th></tr>
<tr><td colspan="2">Perca fluviatilis<br>Gobio fluviatilis<br>Leuciscus Erythrophthalmus<br>Cobitis barbatula<br>Esox Lucius<br>Salmo Trutta<br>Salmo Umbla<br>Anguilla acutirostris<br>(8)</td></tr>
<tr><th>Lough Neagh (North) only</th><th>Lake Geneva (South) only</th></tr>
<tr><td>Gasterosteus aculeatus + G. pungitius<br>Abramis Brama<br>Salmo Salar + S. Eriox, S. Fario, S. ferox<br>Coregonus pollan<br>Anguilla latirostris + A. mediorostris<br>Petromyzon marinus + P. fluviatilis, P. Planeri<br>Ammocœtes branchialis<br>(14)</td><td>Cottus gobio<br>Cyprinus carpio, + C. tinca, C. jesses, C. rutilus, C. alburnus, C. jaculus, C. bipunctatus, C. phoxinus<br>Coregonus thymallus + C. fera, C. hiemalis<br>Lota vulgaris<br>(13)</td></tr>
</table>

In short—as is the purpose of a summa—Thompson is at the forefront of what we now call biogeographical understanding of habitat populations. *Contra* Allen, his description of Ireland's natural history exhibits an extraordinary completeness and *ecological* value that transcends the genres of the national geography travelogue and the specialist sporting manual. There is nothing of the island's local beauties and curiosities (including entertaining anecdotes) for leisured visitors, nor advice on where to hunt, shoot, fish, or collect Ireland's common or rarer species. Instead, Thompson's study recasts and understands Ireland's overtly confluent 'Continental-Hibernian' natural history following, and indeed directly translating, the latest continental science models. Foremost was George Cuvier's *Règne animal* [Animal Kingdom] (1817, revised second edition, 1828). Foundational in its new approach of comparative anatomy, it spawned major national investigations and book-length studies of mammals, birds, invertebrates, etc. in the 1830s and 1840s, including Cuvier's definitive twenty-two volume *Histoire naturelle des poissons* [The Natural History of Fishes] (1828–49), co-authored with Achilles Valenciennes (1794–1865, who completed it after Cuvier died in 1832). The Fishes section of volume 4 of Thompson's *Natural History of Ireland* in 200 pages is a no less massive or ambitious endeavour for world ichthyology. Its significance, like that of its Belfast-based author, remains hidden in full view if regarded only in regional and/or secondary Anglophone-/Anglo-centric contexts. Is there evidence for its foundational value in confluent 'Continental-Hibernian' natural history?

## 'The Leading (Belfast) Naturalist of his Day': 'M. W. Thompson de Belfast'

The history of nineteenth-century European ichthyology triggered by the discipline-defining *Histoire naturelle des poissons* by Cuvier and Valenciennes has yet to be written, and in the form of a companion to Cuvier's extensive survey of all prior knowledge of the field in its first volume (1828). All twenty-two volumes richly attest to the European and international network of expert correspondents, including women, that Cuvier (and Valenciennes) relied upon and quoted to provide new information and specimens to complete this encyclopaedic account and reclassification of world fishes. I unpacked Cuvier's internationally collaborative methods as also the model for understanding such a history of nineteenth-century ichthyology when I mapped one of its 'affluent' branches—the history of 'British' freshwater ichthyology from 1800 to 1870—to recuperate and embed the hitherto invisible *Fresh-Water Fishes of Great Britain* (1828–38) by Sarah Bowdich (1791–1856).[16]

16 Mary Orr, '"Fish with a Different Angle": *The Fresh-Water Fishes of Great Britain* by Mrs Sarah Bowdich (1791–1856)', *Annals of Science*, lxxi (2014), pp. 206–40.

In Table 8.2 are the major Anglophone contributors to 'British' ichthyology I examined in this study, Thompson included, whose publications all failed to acknowledge her work. By now focusing on Thompson in this roster of key book-length publications in the chronological order of their appearance, I deploy his 'Lough Neagh and Lake Geneva' comparative bio-geographical approach overtly to include French/Francophone texts of this period—marked in bold—since French rather than English was the vernacular *lingua franca* of nineteenth-century science and hence of ichthyology. By 1856, William Thompson and his work appear secondary, derivative, and belated, at best regional newcomers albeit expert in an already established (Great) British field. Even the *Dictionary of Irish Biography* entry for Thompson indicates his tributary *assistant* status to greater others within Anglosphere and continental contexts:

> Thompson became the most important naturalist in mid nineteenth-century Ireland. From **1827 to** 1852 he contributed **almost eighty papers on Irish natural history** to the *Magazine of Zoology and Botany* and the *Proceedings of the Zoological Society of London*. From **1836 to** 1851 he contributed to *The Magazine of Natural History*. Invited to travel to the Levant and the Aegean Sea in April–July **1841** with Edward Forbes, professor of natural history at the University of Edinburgh, on HMS *Beacon*, Thompson observed twenty-three species of birds on migratory flights, and published 'Notice of migratory birds' in *Annals of Natural History*. ... Thompson published other papers in the same journal during **1841–3**. At a meeting of the British Association in Glasgow in 1840 his 'Report on the fauna of Ireland—Division Vertebrata' attracted favourable notice. He presented and published a second and final part enumerating the invertebrates at the Cork meeting ... in August 1843. The two reports formed the most complete catalogue of Irish fauna yet published. ... He was president of the Belfast Literary Society (1837–9) and also an enthusiastic patron of the visual arts in the city. ... Thompson was a corresponding member of natural history societies in Boston and Philadelphia and had many friends; he is known to have assisted many other researchers in Ireland, Britain and the Continent. One of those who thought highly of his work was Charles Darwin.[17]

Within the Francophone comparative contexts that frame all the protagonists/works in Table 8.2, however, Thompson's more significant and *leading*

17 A. O'Brien and L. Lunney, 'William Thompson' in *Dictionary of Irish Biography*: https://dib.cambridge.org.

**Table 8.2 Contributors to 'British' Ichthyology, 1828–68**

| |
|---|
| Sarah Bowdich, *The Fresh-Water Fishes of Great Britain* (London: Ackerman, 1828–38) |
| **Georges Cuvier and Achilles Valenciennes, *Histoire naturelle des poissons* (Paris: F. G. Levrault, 1828–49)** |
| William Jardine, *Ichthyology: Fishes of the Perch Family*, The Naturalist's Library (Edinburgh: W. H. Lizars, 1835) |
| Leonard Jenyns, *Manual of British Vertebrate Animals: or Descriptions of all the Animals Belonging to the Classes, MAMMALIA, AVES, REPTILIA, AMPHIBIA, AND PISCES* (Cambridge: John Smith, 1835) |
| William Yarrell, *A History of British Fishes* (2 vols., London: John van Voorst, 1836) |
| **J.-N. Vallot, *Ichthyologie française, ou histoire naturelle des poissons d'eau douce de la France* (Dijon: Imprimerie E. Frantin, 1837)** |
| James Wilson, *An Introduction to the Natural History of Fishes: being the Article 'Ichthyology' from the Seventh Edition of the* Encyclopaedia Britannica; *with above One Hundred and Thirty Illustrations* (Edinburgh: Adam & Charles Black; London: Simpkin, Marshall and Co. and others, 1838) |
| William Swainson, *On the Natural History and Classification of Fishes, Amphibians and Reptiles*, The Cabinet Cyclopaedia (2 vols., London: Longman, Orme, Brown, Green and Longmans; John Taylor, 1838–9) |
| William Jardine, *Illustrations of British Salmonidae*, 2 Parts, Printed for the Author (Edinburgh, 1839–41) |
| **Louis Agassiz, *Histoire naturelle des poissons d'eau douce de l'Europe central* (Neufchâtel: H. Nicolet, 1839)** |
| Robert Hamilton, *Ichthyology: British Fishes*. Parts 1 and 2, The Naturalist's Library (Edinburgh: W. H. Lizars, 1843) |
| William Thompson, *The Natural History of Ireland*, vol. 4 (London: Henry G. Bohn, 1856) |
| H. Cholmondeley Pennell, *The Angler-Naturalist: A Popular History of British Fresh-Water Fish. With a Plain Explanation of the Rudiments of Ichthyology* (London: John Van Voorst, 1863) |
| Jonathan Couch, *A History of the Fishes of the British Islands* (4 vols., London: Groombridge, 1867–8) |

contributions only emerge when we read the facts of his biography—marked in bold above—overtly for his many *continental* activities and correspondents, featuring as unexplained asides in the *Dictionary of Irish Biography*. A reason why 1827 and then 1836 were turning points for Thompson's prolific contribution to scientific print was his extensive non-British scientific travelling in 1826 and again in 1835. His visit in 1826 to Holland, travelling down the Rhine to Switzerland, on to Rome and Naples before returning to Belfast via

Florence, Geneva, and Paris was to view more than the art and monuments of European culture. In 1826, the Paris Museum of Natural History [Muséum national d'histoire naturelle] was the foremost European scientific centre for natural science research, for the collection and classification of new species from voyages of discovery to all regions of the globe, and hence the magnet for European and overseas collaborators and correspondents with the professors at the Paris Museum, such as Cuvier. Coenraad Jacob Temminck (1778–1858), the foremost Dutch authority on Birds and the first director of Leiden's equivalent institution, was a regular visitor. Because science was not yet 'professionalized', expert overseas amateurs were especially welcomed to the collections and laboratories. In the fishes section of *Natural History of Ireland*, Thompson quotes extensively, and without English translation (which still does not exist), from the Cuvier–Valenciennes *Histoire naturelle des poissons*.[18] Its prospectus, which Cuvier published in 1826, promoted the most extensive and intensive collaborative project in this field yet to be mounted. Among the major areas of knowledge he identified as hitherto under-researched were freshwater, and deeper sea fishes in every continent, including Europe.

If 'Paris' thus had profound reflective and inter-connectively continental scientific value for Thompson, so too had 'Geneva' and 'Switzerland', as the title of his final appendix underscores. Their importance for ichthyology in the 1830s and 1840s lies in the scientific person behind the geographical place, Louis Agassiz (1807–73). In 1828–9, Agassiz had corresponded with Cuvier concerning Spix's study of Brazilian fishes in the misplaced hope of joining Alexander von Humboldt's Asiatic Expedition to work as an expert in this branch of natural science. Agassiz then went to Paris in 1830–2 to work instead with Cuvier, until the latter's death cut short Agassiz's opportunities to develop his expertise further in fossil fishes as well as living species. Humboldt then intervened in Agassiz's mounting financial difficulties in Paris by securing him a Professorship back in Switzerland at Neuchâtel. There Agassiz launched his work on the freshwater fishes of central Europe, and their connections with glaciation. When the initial number of his 'Poissons fossiles' [The Fossil Fishes] was published in 1834, Agassiz received his first invitation to England to speak to the London Geological Society, and to receive its Wollaston Medal. The people he would meet included William Buckland (1784–1856)—whom Agassiz later invited to Neuchâtel in 1837—Sir Roderick Murchison (1792–1871), and Sir William Jardine (1800–74). The 'continental' gloss on these 'Anglosphere' facts is that Agassiz spoke no English before he emigrated

18 See, for example, *Natural History of Ireland*, iv. pp. 84, 100, 102, 113. The references to 'Cuv. and Val.' as authority for comparison and the source of agreed nomenclature for the fish under discussion are too numerous to list here. Thompson also cites in French from Cuvier's second edition of the *Le règne animal* (Paris, 1828), pp. 103, 187, 217, 220.

to the USA in 1848,[19] so communicated his scientific papers and corresponded in French. The same situation pertains in 1835, when Agassiz attended the meeting of the British Association for the Advancement of Science in Dublin. There if not already also in London he would have met the renowned collector of fossil fishes, Lord William Willoughby Cole, Third Earl of Enniskillen (1807–86),[20] as well as William Thompson, the now highly respected member of the British Association for the Advancement of Science (BAAS). Because glaciers and fossil fishes were then central to Agassiz's work from 1835 to 1840, he also attended the Glasgow BAAS meeting in September 1840 as part of a three-month research trip with Murchison to the north of Scotland, as well as to Ireland and northern England to inspect Great Britain's Old Red Sandstones and their fossil fishes. The Professor of Natural Philosophy at Edinburgh University, James D. Forbes (1809–68), Sir William Jardine, Lord Cole, and William Thompson all entertained him during this visit, and Agassiz also encountered the work if not also the person of William Yarrell (1784–1856), Secretary of the Zoological Society of London, 1836–8, and expert on British fishes. The making of the biogeological and biogeographical sciences of the 'Fishes of Lough Neagh and Lake Geneva' thus interconnect and inter-reflect their major makers in the 1830s and 1840s—Thompson and Agassiz—as 'affluent' and 'confluent' authorities and collaborators in the principal or 'recipient' mainstream of the Cuvier–Valenciennes *Histoire naturelle des poissons*. Indeed, Thompson clearly qualifies the nature of his relationship with Agassiz and its longstanding qualities in *Natural History of Ireland*: 'A great deal might be said on the manifold influences affecting this species [the Char of Ireland], but it is for my friends, the authors of the two great works *now in progress*—M. Agassiz, in his Fresh-water Fishes of Central Europe, and Sir W. Jardine, in his Scottish Salmonidæ—to descant upon them'.[21]

Mainstream 'European' ichthyology then supplies the missing evidence for Thompson's full, primary, and early involvements in its major classificatory fields, rather than the new niche of fossil fishes pioneered by Agassiz. Like Yarrell, Jenyns, Couch, and other specialists in British ichthyology (see Table 8.2) Thompson's interests spanned marine, estuarine, and freshwater species because of Ireland's particularly extensive and geologically distinctive coastlines and inland waterways. Like all his major Anglophone contemporaries

19 J. Marcou, *Life, Letters and Works of Louis Agassiz, with Illustrations*, vol. 1 (New York and London, 1896): 'The most active part of his life, as regards great discoveries, was spent at Neuchâtel ... where French was the only language spoken. Before he came to America, all his correspondence with English naturalists was in French' (p. xiii).

20 See Biodiversity Heritage Library: https://www.biodiversitylibrary.org/bibliography/5557#/summary.

21 *Natural History of Ireland*, iv. p. 167 (emphasis added).

and correspondents, like Jardine and Jenyns, who were expert fishermen, Thompson was additionally among the earliest experts in nineteenth-century ichthyology regularly to employ dredging as a new method to reach seabed or shore-based species hitherto not part of commercial or leisure fishing. His first experience of dredging was in 1834 with the Manx-born naturalist Edward Forbes (1815–54). Thompson then travelled once more to France, Germany, and Switzerland with Forbes in 1835, before joining him again in 1841 to work on explorative dredging in the Mediterranean and in the Aegean for eighteen months on the *Beacon*.[22] If Thompson's work on algae and bird migrations are recognized as major scientific contributions of this expedition, its applications to what was pioneering about the work on the fishes of Ireland has never been noted. Its innovation surfaces through piecing together Thompson's articles, their reworking in volume 4 in *Natural History of Ireland*, and key references to Thompson in various volumes of the Cuvier–Valenciennes *Histoire naturelle des poissons*, which was undertaken in the order of Cuvier's reclassification of fishes by their major families. By 1839 and volume 13, Valenciennes had finally arrived at the last of the labroid fishes, the 'crenilabri', which are varieties of wrasse:

> Pendant que j'imprimais ces recherches sur le *labrus rupestris* de Linné, je reçois par l'obligeance de M. William Thompson, vice-président de la société des sciences naturelles de Belfast, une notice sur les crénilabres d'Irlande[1], extrait de le Magazine de zoologie et de botanique, n°. ii. Je trouve dans ce travail que l'auteur a fort bien reconnu le *goldsinny*, de Jago dans le *labrus rupestris* de Linné, et qu'il en a pris des individus sur les côtes de Bangor tout-à-fait semblables aux nôtres.
>
> [1] Contr[ibution] towards a Knowl[edge] of the crenilabri of Ireland, p. 3.[23]
>
> [As I was publishing this research on the *Labrus rupestris* of Linnaeus, Mr William Thompson, the vice-president of the Natural History Society of Belfast, kindly sent me a note on the Crenilabri of Ireland taken from the Magazine of Zoology and Botany No. ii. I report in the work that the author has clearly recognized Jago's *Goldsinny* in Linnaeus's *Labrus rupestris*, and that he has caught individuals on the Bangor coast unmistakably like our own.]

22 Its other British naturalists were Robert McAndrew, George Barlee, John Gwyn Jeffries, fellow Irishmen Robert Ball, Edmund Getty, and George Crawford Hyndman.

23 George Cuvier and Achille Valenciennes, *Histoire naturelle des poissons*, XIII *Labraoïdes* (Paris, 1839), chap. 4: 'De quelques genres voisins des Crénilabres et en particulier des Cténolabres', p. 231.

Thompson is clearly corresponding with Valenciennes (in French) to share expertise and update him with new knowledge. But Thompson's article, which Valenciennes correctly quotes as published in the second volume rather than number of the *Magazine of Zoology and Botany* article in 1838, is much more substantial than a 'note'. It runs to 8 pages, with two important plates at the end of the volume, one in colour (see Figures 8.1. and 8.2). Valenciennes thus reports, but does not clarify, what is especially significant about Thompson's work, despite its immeasurable value as comparative evidence of the existence of this particular fish in Ireland. The original title is 'Contribution towards a knowledge of the Crenilabri (*Cuv.*) of Ireland, including Descriptions of Species apparently new to Science'. Thompson will have encountered them often in the Mediterranean during his various dredging expeditions, and through reading the work of the Italian ichthyologist and expert on Mediterranean Fishes, Antoine Risso (1777–1845), whom he quotes in the article. What is new for the ichthyology of Ireland, however, is that other varieties of this species had not previously been found there, or found so far north, although Bangor and Strangford offered warmer and more sheltered habitats to fishes otherwise not found in Ireland's waters, including the *Crenilabrus rupestris* or Jago's Goldsinny mentioned by Valenciennes. The *Crenilabrus microstoma* (Figure 8.2) was, however, found, as Thompson reports, 'on the beach of county Antrim near Cairnlough [*sic*] by … Dr J. L. Drummond, when engaged in collecting *Algæ*'.[24] Carnlough is even further north than Bangor. In such carefully logged and detailed findings and dissemination of his research Thompson offers here not only the first mapping of the ranges of particular European-Mediterranean fishes, but also a meticulous record of unexpected *changes* in their 'natural' environments. We now know that 'the Little Ice-Age' gripping Continental Europe until the 1850s is drawing to its close. The warming waters would encourage more northerly extensions to fish populations.

The continental significance of contributions by 'M. Thompson de Belfast' to the *Histoire naturelle des poissons*, however, also extends to 'common' freshwater species, such as the Bream, covered by Valenciennes in 1844 in volume 17. In this instance, the reference is second-hand, although no less important for that, because, through the important work of William Yarrell (1784–1856), Thompson's longstanding friend who was at his bedside in London when Thompson took ill and died in 1852:

> La brème de Buggenhagen doit être peu commune en Angleterre; car je ne la vois figurée que dans le supplément de l'Histoire naturelle des Poissons d'Angleterre par M. Yarell [*sic*], pag. 39. L'individu lui a

24 *Magazine of Zoology and Botany*, ii (1838), p. 446.

8.1 *Crenilabrus multidentatus*
Source: *Magazine of Zoology and Botany*, ii (1838), plate XIII

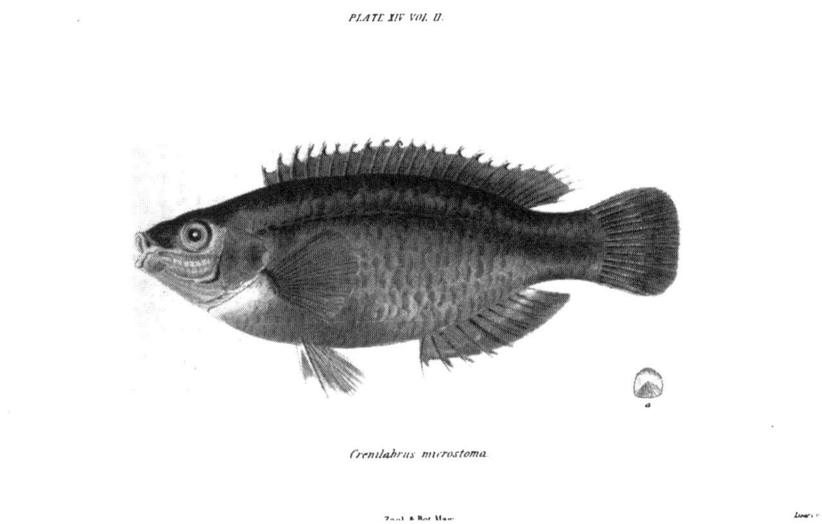

8.2 *Crenilabrus microstoma*
Source: *Magazine of Zoology and Botany*, ii (1838), plate XIV

> été envoyé du comté d'Essex. Cet habile et zélé ichthyologiste nous apprend aussi que M. W. Thompson, de Belfast, à qui l'ichthyologie doit tant de faits importans sur l'histoire des poissons d'Irlande, a aussi rencontré cette brème dans la rivière Logan [*sic*] près de la ville de Belfast. (pp. 56–7)
>
> [Buggenhagen's Bream must be very uncommon in England; for I do not see it illustrated in the supplement to the Natural History of the Fishes of England by M. Yarrell, p. 39. The individual was sent to him from the county of Essex. This skilled and zealous ichthyologist also informs us that Mr Thompson of Belfast, to whom ichthyology owes so many important facts about the history of the fishes of Ireland, has also encountered this bream in the river Lagan near the city of Belfast.]

Thompson is therefore the first to record the uncommon fishes of *England* as well as of Ireland—in this case Buggenhagen's Bream in the river Lagan—that may formerly also have inhabited the larger of the British Isles at the end of the Ice Age. Without thorough expert cross-referencing and cataloguing on Thompson's part, however, disseminated first to Yarrell and other specialists in British ichthyology, Valenciennes could never have recorded the wider significance of this Bream outside its Continental mainland range.

In both instances, Thompson is therefore a leading (mainstream) authority known among his expert contemporaries and without need of intermediaries. A final reference to Thompson in volume 21 of the *Histoire naturelle des poissons* (1848)—in 'ch. viii Des Corégones' [The Coregoni]—fully makes the point and returns us to the species of *Coregonus* found in Lough Neagh and Lake Geneva marked in bold in Table 8.1.

> Le Pollan (*Coregonus Pollan*, Thompson).
>
> M. Thompson m'a envoyé de Loug-Neagh [*sic*], en Irlande la Corégone pollan. ... Les ichthyologistes anglais sont d'accord pour distinguer ce poisson du Gwyniad.
>
> [The Pollan (*Coregonus Pollan*, Thompson). Mr Thompson has sent me from Lough Neagh in Ireland the *Coregonus pollan*. ... English ichthyologists agree in distinguishing this fish from the *gwyniad*].[25]

If the uniqueness of the pollan for Valenciennes is underscored through its comparison with the (Welsh) gwyniad, and through the collective authority of *English* ichthyologists, the French text provides irrefutable evidence of how Valenciennes could make his scientific judgement. Thompson had sent a specimen of this rare 'Hibernian' fish to the Paris Museum for Valenciennes's attention, and to enrich its collections. As noted above in the final appendix of *Natural History of Ireland*, Thompson emphasized the much greater richness of the *Salmonidæ* in the 'northern lake', that is Lough Neagh, than in Lake Geneva (and central Europe). In other words, Thompson's unsurpassed scientific collection of salmonid species in Ireland outstrips any comparable European collection. Species common to both 'Lough Neagh' and 'Lake Geneva' include the *Salmo umbla* or char, now extinct in Ireland, and various species of trout, of which the famed Irish 'gillaroo' (Irish for 'red fellow', due to its distinctive colouring) is historically included as a kind of *Salmo trutta*. Its specific diet of creatures such as snails or crustaceans with shells, which build up in its digestive system, led the Keeper of Zoology at the Natural History Museum, Albert Günther (1830–1914), to rename it *Salmo stomachicus* for this reason. In 1835, however, Sir William Jardine had already defined its voracious habits with the name *Salmo ferox*, since he had also found the variety in Scotland. In the 1830s, classifications of the *Salmonidæ*—salmon and trout—were the most debated by European ichthyologists, including Agassiz, because he and others were beginning to understand their complex life cycle and that the parr was the young of the salmon rather than a separate species.

25 Cuvier and Valenciennes, *Histoire naturelle des poissons*, XXI (Paris, 1848), pp. 502–3.

In the lists in Table 8.1, however, the *Coregoni* or white fishes of the salmon family are of even greater importance, precisely because they inhabit glacial meltwaters and cold mountain lakes and hence indicate Europe's shared post-glacial biogeography. Building on Thompson's already published specialist articles, more than six pages of volume 4 of *Natural History of Ireland* are devoted to its *Coregoni*, although '*Coregonus pollan*, Thompson' is not among the species in the penultimate appendix listing the species that Thompson named. Valenciennes is unequivocal, however, in his reference and use of binomial qualifier 'Thompson', and that '*English* ichthyologists agree in distinguishing this fish from the Gwyniad'. Yarrell and Sarah Bowdich had documented the latter in small lake habitats in North Wales. Jardine had also taken a particular interest in the *Coregonus* of Scotland, the vendace (*Coregonus albula*), in Loch Maben and Castle Loch on his own lands in Dumfriesshire, as these lakes also formed at the end of the last Ice Age. Valenciennes's final reference to William Thompson in what remained the definitive scientific work on fishes throughout the nineteenth century thus fully records the continentally important marine and freshwater species of Ireland as centrally part of a now much less rich nineteenth-century European mainland in terms of certain species. The international authority of the (French) *Histoire naturelle des poissons* and scientific language of Latin binomial naming stereophonically endorse Thompson's significant place as respected expert in European ichthyology. When read as a locational network identifier, the particularizing and potentially patronizing qualifier 'de Belfast' [from/of Belfast] make Thompson and 'Ireland' distinguishing co-contributors, not outliers in ichthyology.

## Fellow-Citizen Science

In its unprecedented, indeed unsurpassed, subject coverage of the fishes of Ireland, Thompson's *Natural History of Ireland* provides ample evidence to determine a response to the rhetorical question heading this essay. The scientific authority of writer and work is unquestionably 'Continental'—international and mainstream—because Thompson's quietly provocative comparative thrust for 'Ireland' in the understanding of its natural history is always as an informatively inter-connective, interdependent, anti-nationalist, endeavour. At the same time, however, the subject coverage of Thompson's work on the fishes of Ireland would not have been possible without his equally extensive Ireland-wide networks garnering expert 'local' knowledge. In these Thompson emulates, but also develops, the multi-participant model with lead orchestrator(s) that Cuvier and Valenciennes pioneered in their *Histoire naturelle des poissons*. We saw above how they meticulously checked and referenced the names and work of contributor-experts such as Thompson to knowledge of given species. Thompson likewise (and unlike Anglophone contemporaries and friends such as Yarrell) assiduously documents the names and contributions of all who collected and sent him specimens and information from every corner and aquatic habitat of Ireland. If his verification of their findings often included comparative consultation of the work of fellow ichthyologists of renown in Europe and Britain, such as Cuvier and Valenciennes, Yarrell or Jenyns, Thompson drew concertedly on wider expertise representing all classes of knowledgeable amateur. Much more than in any other work in Table 8.2, Thompson's 200-page section on the fishes of Ireland contains a plethora of references to *fellow* (specifically Irish) experts, including women, cited for the (scientific) reliability and quality of their information, whether regional, habitat-specific, of a species, or thanks to livelihood/occupation. Information on sticklebacks provides a particularly rich indicative example of Thompson's reliance on his fellow co-workers in the field (of ichthyology):

> March 20, 1835—On examination of a number of 3-spined Sticklebacks from the island of Rathlin (sent by Mrs Gage to Dr. J. D. Marshall, who submitted them to my inspection). I find that in some the lateral plates extend through the entire sides ... No other difference can be perceived in these specimens, which are all of a small size, from an inch to an inch and a half in length. From between tide-marks in Larne Lough (Mrs Patterson); from oozy and rocky pools over which the tide regularly flows, situated near the edge of Belfast Bay (Richard Lantry, Esq.—W.T.); also from a deep pool in the middle of it (Mr James Nichol); and from the harbour of Donaghadee (Capt. Fayrer, R.N.), —I possess examples of the full-armed stickleback of various sizes up to 3 inches.[26]

One therefore finds in Thompson's text frequent references to other well-known Irish experts in ichthyology such as 'Dr. (R.) Ball of Youghal',[27] alongside the equally knowledgeable and zealous but unpublished work undertaken by a cross-section of Irish society, for example a 'Mr W. Andrews of Dingle Bay (p. 82); Rev. G. M. Black (p. 91); 'Major Walker of the Lodge, Enniscorthy' (p. 91); Dr. R. J. Burkitt of Waterford' (p. 91). In footnotes, Thompson also records the equally invaluable, anonymous, expertise of Belfast and Dublin fish-market traders. Because of their livelihoods, Lough and shore fishermen brought insider 'lore' of all kinds that enhanced the subject-defining details of his study:

26 *Natural History of Ireland*, iv. p. 83.
27 See, for example, ibid., iv. pp. 70, 74, 82, 91, 92, 93ff.

> In the months of July and August, 1850, mackerel were unusually abundant on the North-East coast. ... On the 24th August, 26 boats were employed in fishing between Greypoint and Crawfordsburn (Belfast Bay) ... One of the fishermen at Newcastle (James Hill) informed me that 4 men in his boat took 100 dozen in 8 hours.[28]

Thompson is thus a foremost exponent from the 1830s of the collaborative powers of what is currently heralded 'citizen science'.[29] As illustrated in these cameo reports above, however, his altogether more inclusive approach to informed 'non-expert' participants fundamentally challenges modern understandings of what makes citizen science expertise 'scientific', namely that its authority ultimately lies with the 'professional' and with professionalized science. If Thompson verified others' knowledge through his own eyewitness presence in different field sites in Ireland, he set his 'science' always on a par with expertise deriving from longstanding, particular, local knowledge that he could not himself otherwise gain. His overt valorization of entrusted informants from all echelons and educational backgrounds in Irish society ultimately puts scientific endeavour in Ireland on the map as a *foremost* community for the advancement of science. Such a cross-sectional means of collaborative networking is therefore Thompson's model for gauging success in natural history endeavour, both when a new subject or sub-discipline domain is in its scientific infancy—as was ichthyology in the 1840s—and if it is to remain 'leading-edge'. His multi-field methodologies coordinate veritably comprehensive contributions to the sciences of fishes in the 1850s because they channel 'local', 'international', and *cross-society* fellow-practitioner zeal and involvements into a larger common purpose. Without his collaborations with Patterson and other close friends similarly dedicated to the same new comparative scientific standards and excitements, Thompson's endeavour to map Ireland's continental ichthyology could never have come to completion and posthumous publication in the fourth volume of his *Natural History of Ireland*. The openly 'unfinished' nature of Thompson's comprehensive survey and findings on Ireland's aquatic biospheres, epitomized in his final appendix on the 'Fishes of Lough Neagh and Lake Geneva', invites multiple expert involvements of fellow-citizens *for* science today to continue and improve upon his subject-defining work. Such an approach only the more strikingly

28 Ibid., iv. p. 93.

29 For its parameters, see Alan Irwin, *Citizen Science: A Study of the People, Expertise, and Sustainable Development* (London, 1995). For a recent discussion of citizen science in Ireland, see A. Donnelly, O. Crowe, E. Regan, S. Begley, and A. Caffarra, 'The Role of Citizen Science in Monitoring Biodiversity in Ireland', *International Journal of Biometeorology*, lviii (2014), pp. 1237–49. Here, as in most studies, 'citizen science' is a phenomenon dated to 1900.

challenges current marine and freshwater science to re-engage with 'local' and 'cross-border' fellow-citizen experts with distinguishing comparative knowledge of aquatic ecospheres.

## Conclusions

As this essay has everywhere demonstrated, William Thompson's *Natural History of Ireland* was no belated footnote, or appendix, to nineteenth-century scientific knowledge in its period and since. When recontextualized within its contemporary scientific definitions for large river systems, and the new sciences of their 'finny' inhabitants, the 'Fishes of Ireland' section proves a precedent-setting work for the undertaking of country-specific natural history, both in its depth and comparative breadth of specialist investigation, and in its engagement with and contributions to discipline-defining international scientific frameworks. As the final appendix so amply demonstrated as indicative of the broader rationale for *Natural History of Ireland* as a whole, Thompson's comparative anatomy of 'Fishes of Lough Neagh and Lake Geneva' highlights and renegotiates the complex *politics*, as well as the *realpolitik*, of any enduringly important natural history work.[30] Where such endeavour remains determined by overly nationalizing, nationalistic, and 'insular' parameters of engagement, it automatically risks losing its many points of comparative interconnection. Similarly, if the many persons behind scientific knowledge-making are elided or ignored in favour of one or two primary agents, authorities, or professional experts, the history of natural history also loses the greater diversity and community of its networks of players, and the future vibrancy of its development. By contrast, Thompson everywhere shows how much greater is the interconnected flow of knowledge when the significant, and distinctive, fauna of Ireland—common and rare—inform their larger *European*, including British, geo-biological whole, so that this further showcases the astonishing diversity of current and former species in which Ireland remained surprisingly rich in Thompson's time.

The interdependency of Thompson's expertise—his fluencies in and connections with 'French' and 'British' international natural science, his field-practitioner experience in multiple aquatic habitats in the Mediterranean as well as Ireland—therefore reflect his more open model for Ireland's multiply interdependent positioning with regard to its natural history. Non-migratory and migratory species and their ranges—birds, algae, terrestrial and aquatic fauna—are affluent and confluent (Zornlin's terms) in the longer histories

30 As indicative of the politics and *realpolitik* of citizen science, see J. P. Cohn, 'Citizen Science: Can Volunteers Do Real Research?' *BioScience*, lviii (2008), pp. 192–7.

and changing geographies of Great British and Continental European natural history. Today, and especially in a post-Brexit future, Thompson's ichthyology for the whole of Ireland thus offers a comparative bio-geography that locates factors such as extension or contraction of species ranges—migrations—and their study that nuance and recalibrate understandings of 'Irish', 'British', and 'European' marine and freshwater habitats as shaped by geological, geopolitical, as well as Anthropocene 'climate' change, especially since the mid-nineteenth century.

In the 'Fishes of Ireland' covered in volume 4 of his *Natural History of Ireland*, Thompson therefore leaves an even more important critical legacy. This is captured in the fluvial contexts in which they were studied, geologically as well as metaphorically. Commentators and historians of science applying insular Anglosphere and Anglophone lenses, and locked into modern-day understanding of science disciplines, inevitably miss, or marginalize Thompson and his work by regionalizing him (merely) as an Irish naturalist. By contrast, this essay has everywhere underscored the polymathic and *bio*-connected nature of Thompson's work as affluent and confluent with others. Thompson thus reminds all who research on nature and the environment in nineteenth-century Ireland that essentially one needs to know who read, knew, and worked with whom, and in several European language contexts. Only *comparative* intellectual and scientific backdrops permit 'Ireland' to find the continental resonances of its particularly fluid national and international borders, and to disclose as 'Hibernian' certain species and protagonists such as Thompson that remain unprecedented in all other regions of the globe.

If Thompson's place in international reference works, such as the multiple-volume *Histoire naturelle des poissons* of Cuvier and Valenciennes, measures his greater European-Hibernian significance, similar benchmarks can be applied to the reinvestigation of subsequent studies of Irish and British and natural history, including Praeger's. But Thompson's innate understanding of the multiple *grounds* for his research surpasses that of even Cuvier and Agassiz. Because Thompson worked intimately with the habitats of his research subjects, particularly on his dredging expeditions, he was particularly sensitive to complex and changing 'ecospheres', and to the vital importance of mapping them for their longer economic, cultural, and scientific *value*. His work on fishes is therefore distinctive, and with hindsight potentially more leading edge than either Agassiz's—on fossil fishes—or the now forgotten monument of laboratory classification, Cuvier's *Histoire naturelle des poissons*, because living fishes (of Ireland) integrally map what is often invisible or goes unnoticed in changing terrestrial landscapes.

In recuperating Thompson, this essay therefore eschews revisionist reversals of colonial–postcolonial power oppositions to promote a different approach to natural history 'from below' or from the 'periphery' that has nothing to do with 'subalterns' speaking at last.[31] Rather, contextually appropriate river definitions offer more connective and useful critical terminologies for rethinking how knowledge of aquatic forms revitalizes the gamut of human terrestrial knowledge. In addressing 'the question of "Hibernian" fishes', this essay on Thompson's central contributions to ichthyology, and to Ireland's central status in nineteenth-century natural history, inspires new 'fluvial' impetus for the undertaking of more inter-culturally informed nineteenth-century history and history of science. The 'modest proposal' on which it closes and which it opens up is that the use of overly territorial, and terrestrial, biogeographical narratives and metaphors—including the recuperation of 'gaps' and 'blank spaces'—is self-defeating in the study of nineteenth-century colonial history and geography of Ireland. Following the multiple cues for what is unequivocally mainstream about Thompson's work on Ireland—including his reliance on a plethora of scientific contributor-*con*tributaries[32]—it is time instead to follow up on schools of knowledge that plumb more hidden depths, pools, and confluences of intercultural scientific investigation. 'Green' issues are only truly green for Ireland, and the wider European Continent, when its larger inland waters, and fluvial outlets to its seas, reconnect its geology and natural history with international (nineteenth-century) human networks that determine a richer as well as more inclusive scientific understanding and naming of things.

31 See G. Chakravorty Spivak and R. C. Morris, *Can the Subaltern Speak? Reflections on the History of an Idea* (New York, 2010).

32 I am very grateful to David Brown for flagging in his reading of an earlier version of this chapter the importance, and hence re-categorization, of Thompson's co-workers if they were not then to become 'tributaries'.

# 9

# The Ocean of Truth: Atlantic Imagery in Emily Lawless's *Major Lawrence, F.L.S.* (1885) and *Grania: The Story of an Island* (1892)

*Patrick Maume*

Emily Lawless (1845–1913) draws increasing attention from students of nineteenth-century Irish women's fiction, and Irish cultural history more generally,[1] but much remains to be said about the preoccupations which drive and structure her work and how these reflected tensions and uncertainties of the British and Irish societies she inhabited. This chapter develops Heidi Hansson's argument that the tension between science and art in Lawless's work shows her 'writing in the interspace', revealing apparently fixed identities as illusory by-products of ceaseless fluidity (a very Darwinian concept, calling into question Christian and Aristotelean views of nature as possessing underlying fixed structures accessible to reason).[2] It discusses Lawless's quandary between the anti-teleological world view of classical Darwinism and human search for

1 Recent secondary literature includes: H. Hansson, *Emily Lawless, 1845–1913: Writing the Interspace* (Cork, 2007); H. Hansson, 'Nature, Education and Liberty in *The Book of Gilly* by Emily Lawless', in A. Pilz and W. Standlee, eds., *Irish Women's Writing, 1878–1922: Advancing the Cause of Liberty* (Manchester, 2016), pp. 49–64; P. Maume, 'Emily Lawless's *Maelcho* and the Crisis of the Imperial Romance', *Eire-Ireland*, xli (Fall/Winter 2006), pp. 245–66; J. Belanger 'The Desire of the West: The Aran Islands and Irish Identity in *Grania*', in G. Hooper and L. Litvack, eds., *Ireland in the Nineteenth Century: Regional Identity* (Dublin, 2000), pp. 95–107; G. Meaney, 'Decadence, Degeneration and Revolting Aesthetics: The Fiction of Emily Lawless and Katherine Cecil Thurston', *Colby Quarterly*, xxxvi (June 2000), pp. 157–75; L. Mills, 'Forging History: Emily Lawless's *With Essex in Ireland*', ibid., 132–44; R. Lynch, *The Kirwans of Castlehacket* (Dublin, 2006), pp. 147–65.

2 Hansson, *Lawless*, pp. 58–63, 161–4.

meaning through science, art, love, and religion. It suggests Lawless's stories of Ireland share deep concerns with those set elsewhere, which are comparatively neglected. (*Major Lawrence, F.L.S.* is usually overlooked by critics, while the numerous readings of *Grania* show insufficient attention to the oceanic forces which dominate the novel's Inishmaan and the life of the heroine.)

Lawless was a landlord's daughter and sister (her father, the third Baron Cloncurry, was succeeded by two of her brothers) living through agrarian unrest, suspecting she was born too late for an imagined time when tenant–landlord relations were 'naturally' harmonious; she was a Unionist painfully aware of the violence of conquest and colonization, but convinced nationalism provided no coherent political alternative; a troubled agnostic haunted by Christian faith lost through Darwinism and refusal to accept that suicidal relatives were damned; a woman of ambiguous sexuality. Lawless was brought up near the Atlantic coast of Galway by her mother's relatives the Kirwans of Castlehacket, but spent her later years in Surrey, remarking that she left Atlantic waves to cultivate an English garden.[3] *A Garden Diary* (1909) displays the garden as miniature replica of the natural world—both need rigorous pruning to forestall chaos—and as haunted by bloody struggles to maintain British power in China and South Africa; reflecting her position among 'the mere irresponsible camp-followers of science' contrasted with geniuses ('Darwin played the tune … the rest of us … danced to his piping')[4] as reading narratives of scientific explorers ('with the prince of them all one starts once more upon that immortal *Voyage of the Beagle*')[5] palely reflects the experiences of discovery. This anticlimactic descent from heroic dreams of youth to awareness of her scientific limitations and of the defeat of her Irish landlord class through moral failures long preceding her birth (including failure to preserve tenants from the Great Famine of 1845–9, but perhaps inextricable from the amorality of conquest) underlies Lawless's preoccupation with contrasting naive protagonist and detached omniscient narrator, sometimes employing scientific terminology to describe experiences in which the former—often the latter also—is uncomprehendingly immersed.

Lawless deploys irony at her own expense; *Major John Lawrence, F.L.S.* (1885) features a rich, handsome, selfish, cruel, and tubercular young aesthete called Algernon Cathers, whose grandfather became wealthy by manufacturing blankets.[6] The first Lord Cloncurry made the Lawless family fortunes in this manner before buying his peerage through political corruption, as

3 Emily Lawless, *A Garden Diary* (London, 1901), pp. 8–9, 169.

4 Ibid., pp. 220, 179.

5 Ibid., pp. 3–4.

6 Emily Lawless, *Major Lawrence, F.L.S.* (London, 1888), p. 28. Lawless satirizes her ancestors' displacement of the Aylmers from their Kildare estate in *A Colonel of the Empire* (1895); see Hansson, *Lawless*, p. 142.

nationalists recalled when Lawless criticized them. Cathers talks of becoming a peer, saying anyone with enough money can get a peerage.[7]

Lawless is best known for her historical novels (especially *With Essex in Ireland* (1890) and *Maelcho* (1894)), her Land War novel *Hurrish* (1886), and *Grania: The Story of an Island* (1892), but she also had an interest in science and significant expertise. Her youthful collection and study of marine organisms informs several of her writings. Hansson argues that as an adult her contributions to scientific journals and correspondence with Darwin (on his initiative) challenged the established boundary between female collector of specimens and male analyst. While *Garden Journal* negotiated this space, abounding with 'modesty markers' which allowed her to blend—as this feminine genre required—botanical knowledge with 'life writing, spiritual reflections and cautious scientific critique', elsewhere she contested classificatory models established by British botanists, which often embodied norms not justified by the knowledge of Irish flora and fauna which she helped to develop.[8] This reading suggests Lawless's portrayal of the scientific explorations of the less than conventionally masculine hero of *Major John Lawrence, F.L.S.* does not depict simple failure, but expresses frustration at failures of the scientific establishment to acknowledge anomalous data and by extension the inability of the scientific mind to provide a comprehensive world view by which the ordinary person can live.

Lawless's Darwinism shaped her bleak world view, haunted by lost teleology and the vast geological timespans displacing the relatively compact Mosaic chronology. Lawless also displays suspicion of aesthetes—whose artistic sensibility claims a universal comprehension inaccessible to scientific empiricism—as selfish and self-indulgent; she sometimes combines aesthete and tourist as exemplifying moral superficiality. In *A Millionaire's Cousin* (1885), involving a tour of North Africa, the artist-narrator unconsciously displays the egotism which motivates and wrecks his pursuit of the heroine. In 'Borroughdale of Borroughdale'—*Plain Frances Mowbray and Other Tales* (1889)—the scientist is contrasted with the aesthete as capable of human relationships and social responsibility because he seeks shared truth, not private fantasies. Elsewhere, however, Lawless fears the truth science reveals is inhuman; pleasant illusions may be preferable to life devoid of human meaning. This chapter discusses how these concerns are reflected in the

7 *Lawrence*, 102; J. G. Swift MacNeill, *Titled Corruption* (London, 1894), pp. 64–8.

8 H. Hansson, 'Emily Lawless and Botany as a Foreign Science', *Journal of Literature and Science*, iv (2011), pp. 59–73. See also G. Kehler, 'Gertrude Jekyll and the Late-Victorian Garden Book: Representing Nature–Culture Relations', *Victorian Literature and Culture*, xxxv (2007), p. 622. Page references to *Lawrence* are to the 1888 one-volume edition by John Murray; references to *Grania* are to the 1894 one-volume edition by Smith, Elder and Co.

portrayal of nature and natural history in two of Lawless's novels, *Major John Lawrence, F.L.S.* and *Grania: The Story of an Island*, with particular reference to the Atlantic seen as scientific phenomenon and as symbolizing the vast, unknown, and perhaps unknowable cosmos and the depths of the human unconscious, neither fully captured by scientific classification. The Atlantic also symbolizes, echoing Wagnerian opera, the erotic passion which Victorian social proprieties only allowed Lawless to represent in such terms. In *Grania*, the Atlantic is equated to the peasant heroine's irrational physical attraction to a worthless fiancé, driving her to an oceanic *Liebestod*.

## I

*Major John Lawrence, F.L.S.* is one of Lawless's lesser-known novels, because it does not have an Irish setting and displays some rebarbative features of Victorian sensibility. The hero is a stolid and dutiful Indian Army officer, lame from a childhood accident, who becomes infatuated with a teenage girl:[9] his name recalls the Lawrence brothers, prominent in north-western India in the early Victorian period, and he reflects a heroic stereotype of local Indian colonial administrators selflessly working for the greater good.[10] He is also compared to the hapless William Dobbin of Thackeray's *Vanity Fair*, dominated by nearly hopeless passion for Amelia Sedley who obsessively idealizes her dead husband.[11] Lawrence first meets the heroine Elly Mordaunt when he is thirty-three and she is eleven years old; she is gazing out to sea and he rescues her from the incoming tide.[12] When Elly reaches her late teens and Lawrence realizes he loves her, she—now Lady Eleanor—has fallen passionately in love with—and become engaged to—the loathsome Algernon Cathers. Lawrence senses Cathers's amorality when he first meets him as a small boy, but cannot articulate his misgivings. Cathers, who resembles the heartless aesthete Gilbert Osmond of Henry James's *Portrait of A Lady*, is not a creative artist but lavishes money to display his pretensions as a connoisseur, and thinks Eleanor 'the finest piece of property … he possesses'.[13] Not only dressing in the Aesthetic manner himself, he gets his loving but uncomprehending old mother to adopt the same fashions, for which she is ludicrously unsuited.[14] Lawrence's natural habitat is the Atlantic coast; Cathers inhabits the artificial atmosphere of Mediterranean tourist enclaves, where the local

9 *Lawrence*, pp. 2–3.
10 C. Allen, *Soldier-Sahibs: The Men Who Made the North-West Frontier* (London, 2000).
11 *Lawrence*, p. 352.
12 Ibid., pp. 34–40.
13 Ibid., pp. 148, 266, 305–6.
14 Ibid., pp. 337–8.

population seem only spectators or picturesque objects. (At moments the protagonists sense the wider life around them, as tossing trees of an Italian forest remind Lawrence of Atlantic waves.) Lawless had travelled in the Mediterranean with her mother. After an exquisitely painful marriage, whose later stages in the Mediterranean are witnessed by the long-suffering Lawrence, Cathers finally expires. Eleanor's willingness to stay with Cathers even when he installs his mistress as governess for their children reflects not her concern for the children, whom the governess/mistress mistreats, or pity, but memory of her love even when Cathers has killed the love itself. Lawrence is tormented since Cathers's death provides no escape; the past cannot be changed, Cathers will always have been Eleanor's husband.[15] Lawrence eventually blurts out his love for Eleanor; she reciprocates, but is emotionally unable to consider a second marriage. Lawrence returns to Devon where he finds healing through the Atlantic—every 'hope, ambition, dream of his whole youth seemed to be entangled in that sweep of grey hollows and dancing summits'. His love for the Atlantic is like Elly's for Cathers, though it has not betrayed him as Cathers did Elly—'the old hankering was still alive and keen, keen as first love, not to be obliterated by any length of sober domesticity'.[16] Lawrence finds the Atlantic reawakens memories of the child Elly[17] before he returns to his Indian duties. It is implied Lawrence will return to Devon and a sombre second union with Eleanor, but the anticlimax is reinforced by the reader's sense that Lawrence—by extension the reader—was a partly informed onlooker where Eleanor was the real protagonist.

As a boy convalescing on the Atlantic coast of Devon (which Lawless compares to both Ireland's Western Atlantic coasts and the north of Scotland) Lawrence took to studying marine life, an interest maintained into adulthood.[18] A principal attraction of India is that it shows nature directly and unveiled. In India, Lawrence's studies attract the attention of a Huxley-like professional scientist, Henniker Jenkyll, who secures Lawrence's election as a Fellow of the Linnaean Society. Having inherited a run-down cabin on Devonshire cliffs, Lawrence spends leave periods pursuing his studies:

> Just as a man who is not a great poet may yet enjoy by moments the genuine poetic rapture, so our Major, who was not a great scientist, who had not even had the preliminary training which might have enabled him to become one, had by moments his share of that electric thrill which is the guerdon of those whose names stand written in

15 Ibid., p. 307.
16 Ibid., p. 301.
17 Ibid., p. 302.
18 Ibid., pp. 4–7.

> the Red-letter book of Fame. He dived, he plunged, he floated; he lost breath in waters that were too deep for him; he rose again to the surface, panting, but undefeated; he scoffed at himself, but he persevered; he dreamed, he hoped, he believed. Such at this particular moment of his career was his inside life.

Conveying the excitement of scientific discovery and assimilating it to artistic inspiration, Lawless describes Lawrence's chief preoccupation when he first meets Elly Mordaunt—the pursuit of scientific knowledge, anthropomorphized as a desirable older woman with many suitors but ultimately unattainable:

> He did not avail himself of these [invitations to visit] as much as he might have done. He liked Lady Mordaunt [Elly's grandmother, Lawrence's best friend] immensely, but she had a rival who, had she been a woman, would perhaps by this time have been getting a little into her decadence, but who, being superior to the foibles of humanity, seemed only in her admirer's eyes to get younger and fresher every day ... he had embarked ... upon his old zoological pursuits, and daily despatched pages of closely filled memoranda to his friend Jenkyll ...
>
> Had he not already had his little successes? Had he not discovered no less than three distinct species of Neuroptera, and two of Hymenoptera, one of which had even been endowed with the preposterous name of Laurenceana after its finder? It had certainly been changed again not long afterwards when a fresh revision had been made of the genus, but that, as any reasonable person would admit, was not *his* fault!

Neuroptera and hymenoptera are orders of winged insects, whereas in the novel Lawrence is seen working only on marine organisms. This suggests lack of specialization, while the reclassification suggests the provisional nature of the knowledge he pursues, and its fragility as source of meaning.[19]

> Of late, since his coming to Devonshire, a new light had dawned for him, one of greater brilliancy than any that had as yet risen upon his horizon. He had been working at the life-history of certain undistinguished marine organisms ... fortunately well-represented upon that part of the coast. Suddenly, in the course of these investigations, a new idea had burst upon him—as a seed-vessel bursts when the moment

19 Cf. Hansson, *Lawless*, pp. 58–64 on Lawless's musings about the instability of scientific classification and irresolvable differences between apparently similar organisms.

> for its fruition has come. It was merely an indication; a dazzling flash, and then again obscurity; but it is through such indications, by means of such momentary flashes, that all new and epoch-creating discoveries admittedly have been evolved …

Lawless is using explicitly Darwinian language: these bursts of scientific inspiration are equated with the chance mutations which drive Darwinian evolution. Alas, whether or not Lawrence's amateur theory corresponds to the marine fauna it describes, it is not adapted to the developing ecosystem of late Victorian professional science, and thus destined to perish.

> It was merely a perception of certain affinities, certain underlying similarities which had previously, as it seemed to him, been overlooked … there are two orders, known respectively by the alluring titles of the Ctenophorae and the Discophorae … whose precise modes of development and interdependence have long been a fertile source of zoologic controversy. They have been arranged, divided, and subdivided by one illustrious authority, and then again dis-arranged, re-divided, and re-subdivided upon a totally different system by another … It was in one of the bewildering ramifications of this bewildering group that the Major caught sight of his idea—his *fact*, he called it. He flung himself upon it as a gold-seeker upon a promising vein of auriferous quartz!

Lawless was drawing on real Linnaean controversies. A vigorously argued paper by Alexander Agassiz on the 'Embryology of the Ctenophoræ' described the complexities of these classificatory debates, suggesting Lawless was indeed a 'camp follower' of the natural history debates of her day.[20] Discophoræ is a term (subsequently obsolete) for certain types of jellyfish; the phylum ctenophoræ or comb jellies are invertebrates which swim with cilia. Ctenophoræ had recently been reclassified to distinguish them from the phylum known as colentera. Lawless's description of the excitement of marine research surely derives from youthful experiences as 'a nervous girl' of amateur zoology on the west coast of Ireland. She speaks of catching enthusiasm from Atlantic vastness in language implying she experienced it herself:

> Sometimes at night in the little sitting-room, which by a summary process he had turned into a zoologic laboratory … he would be filled

20 A. Agassiz, 'Embryology of the Ctenophorae', *Memoirs of the American Academy of Arts and Sciences*, new ser., x (Aug. 1874), pp. 357–98. Agassiz was the son of Louis Agassiz, whose career is discussed by Mary Orr elsewhere in this volume.

> with a feeling of wild excitement, which to most men comes only once or twice in their lives, at some great crisis, some culminating moment of their fate. His heart would beat, his nerves tingle, his whole frame shake like a nervous girl's ... his room would suddenly grow too small for him—it was not of commanding proportions—and he would spring to his feet, cross it with hasty steps, and gaze eagerly out westward towards the Atlantic, that Atlantic which to all who have once caught enthusiasm from its vastness, becomes the very symbol and embodiment of greatness, spiritual no less than physical. To John Lawrence, with his head full of his organic problems, it was the symbol of Life itself, that great ocean of sentient life with all its baffling problems, its bewildering and inextricable mysteries. Was there not an Atlantic too in this direction? he would ask himself excitedly—an Atlantic which as yet had hardly been explored, whose skirts and land-locked bays were all that could be said to be known? ... Who could tell at what moment even the least accredited of voyagers might suddenly burst into a new sea, a virgin sea whose waves had never yet been ploughed by mortal keel?[21]

Lawrence eventually realizes he lacks the intellectual tools to develop his theories in a form professional scientists will take seriously, after he approaches Jenkyll (who is too busy with other commitments to test Lawrence's ideas) and another scientist who dismisses Lawrence as an amateur unacquainted with relevant secondary literature. He is advised to write for a scholarly journal but lacks the confidence and terminology—the intimidating complexity of the Agassiz paper suggests he was probably right to be cautious.[22] Lawrence's endeavours are not completely and wilfully futile, as with Mr Casaubon in George Eliot's *Middlemarch*; Lawless implies the scientists are unjust though understandable in their dismissal. Lawrence's tragedy is that he is a latecomer; as a gentleman amateur in the age of Mr Darwin he might have become significant, but he is out of place in the professionalized milieu of Professor Huxley. Moreover, Lawless hints that Lawrence's scientific pursuits are estranging him from humanity. He copes with social visits to his father, who has remarried and fathered young children, by observing family life as 'some new variety of crinoid or Echinoderma'.[23] These hints culminate in a painful revelation about possible human consequences of disinterested pursuit of knowledge from a 'shabby curator' in a local museum:

21 *Lawrence*, pp. 13–17. Lawless's sexualization of the Atlantic ('skirts ... a virgin sea') may be unconscious.

22 Ibid., pp. 69–73.

23 Ibid., p. 55.

'I suppose you know that the Godby Collection is here?' ...

John Lawrence had never even heard of the Godby Collection. 'What is it?' he asked.

'Hydroida—your own subject. Thecaphorae, I take it chiefly—Sertularians, and such like ... Old Godby was quite a specialist ... not a trained man; picked it up by himself. I believe he knew more about the order than any one in England; at any rate, in these parts'. ...

The opening revealed a perfect chaos of specimens, which to an uninitiated eye would have appeared to be seaweeds, crowded for the most part one on top of another, the paper upon which they had been mounted soiled, yellow and torn with age, the writing half obliterated with time and indifferent usage. ...

'Old Godby was forty-five years making it, I believe ... A fellow who knew him at Tenby told me it was his ruin. He was a surgeon, and at one time in decent practice. ... Left his patients to take care of themselves; spent his whole days upon the shore; had to sell his practice; got crippled with rheumatism; parted bit by bit with every stick of furniture he possessed; always stuck to his collection, though,—was adding to it up to the day of his death. ... left it to this museum on condition that it should be known as the Godby Collection! Poor old chap! I suppose he thought that he had secured a sort of posthumous immortality'. The curator stopped, and began licking a piece of delicate brown lacework, fastening it down to the paper like a postage-stamp with his thumb. 'That was seventeen years ago, and, to the best of my belief, you are the first person that has looked at it since!' ...

John Lawrence ... felt indignant and depressed ... 'I should have thought *some one* might have found time to go over the collection,' he said ...

The friendly curator shook his head with an air of deprecation. 'I should have said the same thing myself *once* ... it is only amateurs like yourself—gentlemen with leisure, who can afford to labour, as you say, for the interest of the thing. Once a man gets into a beaten track, he has to wear blinkers. He couldn't do his own work else!' ...

John Lawrence walked away ... The fate of old Godby lay heavy on his soul! Poor old Godby, who had ruined his career; neglected his practice; thrown over his prospects; sacrificed health, fortune, everything. And for what? For an obscure corner in a third-rate local museum, already groaning under the undesired weight of similar donations.[24]

24 Ibid., pp. 92–6.

Once again, Lawless uses contemporary scientific terminology, referring to current debates that highlight the distinction between the amateurs of the early Victorian period and later decades. It does not seem entirely coincidental that she chose hydroida. According to a report J. E. Duerdon read to the Royal Irish Academy in 1893, prior to the surveys of the south-west coast of Ireland in 1885, the only significant work done on the Irish Hydroida was William Thompson's work on Dublin Bay in the 1850s.[25] And yet, the evidence collected by Thompson by means of the 'citizen scientist' methods described by Mary Orr elsewhere in this volume allowed him to assemble a richer account of the prevalence and distribution of Hydroida in Irish coastal waters than Duerdon allowed.[26] Duerdon's judgement perhaps reflects changing attitudes to scientific methods, but it also says something about Lawless's portrayal of Godby. Whether Lawless was conscious of Thompson's intensely collaborative methods cannot be said, but she presents Godby's monomania and social isolation as a warning to Lawrence. In contrast to Lawless's opposition of scientist and aesthete, Godby is presented as a *poète maudit*, destroying himself to pursue a vision he glimpses but can never articulate. To Lawrence—and to the reader—such a life is perhaps wilfully self-indulgent, certainly utterly self-destructive and futile. Lawrence then develops a closer interest in the possibility of more personal fulfilment through Elly Mordaunt.

## II

At first sight *Lawrence* might seem to have little in common with *Grania*, centred on an illiterate young woman living on Inishmaan, the central Aran Island, in the late 1860s. The characters of *Lawrence* move across the globe and live relatively comfortable lives: Grania is confined to an island described as 'the most retrograde spot, probably, within the four seas',[27] and has no conscious desire to leave it because she cannot visualize life elsewhere. The islanders are so tightly constrained by sea and wind that when emerging from their hovels, they seem little more than rocks or fungi embedded in the earth.[28]

*Lawrence* is centred on a male character and the conflicts of Elly/Eleanor

25 J. E. Duerdon, 'Report of the Hydroida Collected by the Royal Irish Academy Survey of the South-West Coast of Ireland, 1885, 1886, and 1888', *Proceedings of the Royal Irish Academy* (1889–1901), iii (1893–6), pp. 137–50. 'Hydroida' is a now obsolete term encompassing certain cnidarian (polyps and medusae) aquatic organisms. 'Sertularians' are also a variety of cnidarian; 'thecaphorae' is a now obsolete term for cryptomedusae.

26 William Thompson, *The Natural History of Ireland* (4 vols., London, 1849–56), iv. pp. 445–60.

27 *Grania*, pp. 33, 104.

28 Ibid., pp. 38, 91–2.

are only glimpsed indirectly; *Grania* is so caught up with its protagonist that it is easy to forget there are scenes where she is not present. In one of these the malevolent storyteller Peggy Dowd compares Grania to a Connemara woman who married into the island (Grania's mother came from Connemara), whose sickly child was identified as a changeling. This woman passionately rejected neighbourly advice to roast the 'changeling' on the griddle or pour a scalding potion over it to secure the return of the 'real' child, and her increasingly frenzied efforts to defend her child were regarded by Peggy (and every islander, including her husband) as 'proof' she was possessed by evil spirits.[29] Lawless cannot have been thinking of the fate of Bridget Cleary (in 1895, three years after *Grania*'s publication), which makes her invocation of the changeling myth even more chilling.

Where *Lawrence*'s Italian scenes and their peasant inhabitants are seen through the detached eye of the tourist, in *Grania* the external observer's gaze is ironized by comparison with the experience of islanders immersed in the landscape and mauled by it. The narrator repeatedly contrasts the outwardly picturesque appearance of characters with their inward suffering and frustration. Grania's father Con O'Malley is seemingly a 'typical' peasant, 'if there are such things as typical peasants, or, indeed, any other varieties of human beings',[30] who is drinking himself to death from grief for his dead wife. Grania's fiancé Murdough Blake, 'ought to have rejoiced the innermost heart of a painter, had a painter ever thought of going to the Aran Isles in search of subjects, a ridiculous supposition, for what painter would ever dream of doing so?'[31] Towards the end of the novel, tourists perceive Grania as a picturesque item of scenery, until she gazes back at them so resentfully that they flee, and when their yacht is beset by a storm she fantasizes that the anger in her blood communicated itself to 'her own wild sea and sky'.[32] As Hansson sees Lawless as challenging colonial fantasies of possession and power by evoking a multivalent and unclassifiable 'interspace', the tourists who saw themselves as detached connoisseurs are caught in the weather systems which define Grania's existence; thus Lawless implicates her readers and herself in Grania's situation.

Lawless's remark that it is unthinkable that Aran should ever attract painters indicates her awkward position on the cusp of the Revival: she recognizes that the inhabitants of Aran resemble the peasants and fishermen who attracted such artists as the French Barbizon School and the habitués of artists' colonies in Brittany, but sees the islands as so remote that it

29 Ibid., pp. 145–55.

30 Ibid., pp. 21–2.

31 Lawless has forgotten, or is unaware of, Frederic William Burton's 1841 watercolour, *The Aran Fisherman's Drowned Child*.

32 *Grania*, pp. 272–4.

never occurs to her that they might actually attract aesthetic interest. Revival writers challenged *Grania*'s claim to represent the authentic inner lives of the islanders as distinct from the tourist view. Synge noted that Lawless had not lived among the islanders as he had. (Ironically, Synge's Darwinian portrayal of tramps as 'variants'—mutations which cannot adapt to their environment—resembles Lawless's attitude to Grania). Lady Gregory thought *Grania* a particularly inauthentic representation of Aran.[33]

*Grania* does bear a relationship to *Lawrence* as the photographic negative is related to its print. They are united by oceanic imagery of immersion; Grania, like Lawrence, is entrapped by social limitations and inarticulacy, but to a much greater extent. Lawless contrasts the familiar sea linking the islands to the coasts of Galway and Clare with 'the Atlantic, a region which, despite the ploughing of innumerable keels, is still given up by the dwellers of those islands to a mystic condition of things unknown to geographers, but too deeply rooted in their consciousness to yield to any mere reports from without'.[34] The islands are virtually swamped by the Atlantic, which it is suggested will eventually submerge them; sea waves are individual soldiers in an army bearing down in unison on an isolated foe (the island).[35] When the rain falls limestone seems to melt underfoot as if the island were dissolving. Conversely, the air, filled with water vapour, seems solid; the wind, scouring the island, even eroding stones, is an 'aerial ocean without bound or limit':[36]

> If the destiny of the islands was sooner or later to be washed away and to vanish from sight in the sea, it seemed as if now was the time that destiny was likely to be fulfilled. The rain came down in lined sheets, and in sheets it swept over the surface. There being no earth for it to dry into, it poured over the level slabs, sweeping from slab to slab almost as the sea swept over the rocks between the tide-marks. Watching it at such moments, it would have seemed to you as if the whole island would shortly become one great waterfall, or scarcely perceptible reef for the Atlantic to roll over.[37]

In the end, Grania, whose physical strength and speed are constantly emphasized, is miserably drowned in the ocean. Hansson's view of Lawless celebrating an 'interspace' uncontainable by colonial power-fantasies perhaps fails to recognize how Lawless's vision entails inescapable vulnerability, terror, and abjection.

33 Murphy, *Lady Gregory: The Journals*, ii. p. 153; Hansson, *Writing the Interspace*, pp. 71–85.
34 *Grania*, p. 2.
35 Ibid., p. 168.
36 Ibid., pp. 37, 46, 170.
37 Ibid., p. 167.

The narrator presents islanders driven by natural forces of which they have only a limited understanding and ignorantly shaped by a deep history which the narrator—who regularly uses scientific terms—and the reader comprehend. They are presented as virtually asexual and ignorant of the depths of their own emotions because confined by island life.

The island flux's defeat of attempts to shape and control it is reflected in the absence of gentry. Two standard literary tropes justifying the Irish landlord class were the attempt of such writers as Charles Lever to portray landlord joined to peasant by paternal feudal personal loyalty cemented by convivial drunkenness[38] and the liberal claim that landlords were necessary agents of social and economic modernization through improvements such as those undertaken by Lord Palmerston.[39] In *Grania*, the only relic of landlord attempts at improvement is bent-grass planted to protect sand dunes from the wind. The grass is no longer supervised or renewed; islanders with cattle—including Grania—graze it, despite knowing this accelerates the erosion of the island.[40] The breakdown of Leveresque fantasy is represented by a ludicrously inappropriate derelict mansion—rapidly eroded by the climate and local vandals—built by a landlord who moved to the island some decades ago to escape imprisonment for debt.[41] This resident landlord is remembered only by the senile Pat Durrane, who mumbles that the gentleman was a good man and those were good times, while failing to explain their alleged goodness and giving ample reason to suspect that the gentleman was hardly good.[42] Lawless's 1886 novel *Hurrish* presents a virtuous peasant mourning the breakdown of paternalism, destroyed by fantasizing agitators; *Grania* still sees fantasy as destructive but implies that paternalism never existed and contentment with the intolerable is stultification. If *Grania* shares the Revivalist concern to articulate and make sense of the inarticulate experience of the peasantry, it was also Counter-Revivalist in its insistence that these inarticulacies are not remains of a coherent and retrievable ancient civilization, but can only be reshaped by the modern individual mind.[43] The supposed Leveresque house of hospitality becomes a squalid, anhedonic

38 P. Maume, 'Music-Hall Unionism: Robert Martin and the Politics of the Stage-Irishman', in P. Gray, ed., *Victoria's Ireland? Irishness and Britishness, 1837–1901* (Dublin, 2004), pp. 69–80.

39 D. Brown, *Palmerston* (New Haven, CT, 2010); P. Maume, 'Works, Righteousness, Philanthropy, and the Market in the Novels of Charlotte Riddell', in Pilz and Standlee, *Irish Women's Writing, 1878–1922*, pp. 17–32.

40 *Grania*, p. 67. Contrast David Brown's description of Palmerston planting bent-grass to halt erosion on pp. 44–45 of this volume.

41 Ibid., pp. 52–3.

42 Ibid., pp. 207–13.

43 P. Maume, *'Life that is Exile': Daniel Corkery and the Search for Irish Ireland* (Belfast, 1993); J. W. Foster, *Fictions of the Irish Literary Revival: A Changeling Art* (Dublin, 1987).

drinking den; Murdough drinks and fantasizes in the deserted villa, whose ludicrous architecture resembles his fantasies of being a gentleman.

The other prominent buildings on the island are a ruined Catholic church and monastic cell (there is no resident priest); Grania's pious half-sister and mother-substitute Honor is associated with these and the rival Catholic dream of an idealized past. While the islanders barely recall their own recent historical past, they are acutely aware of these early Christian ruins, while ignorant that geology and palaeontology show the world to be millions of years older than they imagine. In one striking narrative comment, Lawless imagines the islanders obliviously inhabiting the skeleton of a large maritime dinosaur resembling the fossil plesiosaurus discovered and described in Dorset in 1823–4:

> A curious symmetry marked the first of these steps, up the backbone of which the sisters were then mounting: you would have been struck in a moment by its resemblance to the backbone of some forgotten monster, unknown to geologists. A python, say, or plesiosaurus of undetermined species, but wholly impressive vastness, stretching itself lazily across a third of the island, till its last joint, sinking lazily towards the sea, disappeared from sight in the general mess of loose stones which lay at the bottom of the slope.[44]

There are problems of condescension and impersonation in writing from the outside about people whose consciousness of their situation is limited. Lawless tries to avoid this danger, for example, by replacing conventional 'brogue' dialect (associated with low comedy) with a stilted form of speech conveying that the characters are speaking Irish.[45] Whether Lawless succeeds is another matter; Murdough Blake can be seen both as a recognizable social type and as a colonialist stereotype of the colonized as childish, effeminate, and lacking authentic virility, though in this he resembles Algernon Cathers, hardly a colonial subject. Murdough, addicted to daydreaming about his own imagined greatness and 'borrowing' Grania's earnings to drink, is the islands' nearest approach to an aesthete.[46] Unlike Lawrence and Godby, he never makes concrete efforts to realize his dreams. Lawless suggests that a few decades later he would have made speeches at Land League meetings.[47]

44 *Grania*, pp. 35–6.

45 Ibid., pp. 35–6. In a 1913 letter to Lady Gregory, Lawless admits that 'the dialogue in my poor old Grania was [a convention] … an attempt to escape from the brogue, with all its tiresome and stale associations'. See D. Murphy, ed., *The Journals of Lady Gregory*, II: *Books 30–44* (Gerrards Cross, 1987), p. 419.

46 *Grania*, pp. 286–90.

47 Ibid., p. 61.

Even Grania sometimes recognizes marriage to Murdough promises abject disaster.[48]

*Grania* has attracted attention for its bleak portrayal of domestic abuse, when Grania, returning from a fair in Galway where Murdough gets drunk and lets her down, sees a woman, visibly bruised, with a drunken husband and hungry children, and realizes this is her probable future.[49]

Grania's relationship with Murdough is more complicated, however, than this scene suggests. Analyses which treat it in isolation neglect Grania's intense physical attraction to Murdough, which draws her back repeatedly and disastrously. She sometimes expresses this desire, despite local convention and her own misgivings, to an extent which horrifies Murdough and startles Grania herself. Although one minor male character is passionately fond of his wife, for which he is derided by everyone including the wife,[50] Lawless assumes that because islanders do not express sexual attraction in public, they do not experience it at all.[51]

Although Grania, unlike Murdough and Honor, does not consciously wish to escape the island, her desire for Murdough reflects not only her own physical feelings but yearning for release similar to that which Murdough finds in alcohol and Honor in religion. Grania's sudden eruptions of passion are assimilated by Lawless to the Atlantic.

One of Grania's two most passionate moments comes when she and Murdough are sea-fishing on a summer day. The island fishermen are so assimilated to the sunny ocean that their boats seem sea-creatures rather than artificial constructions, and Lawless's narrative voice intersperses amongst descriptions of Grania and Murdough's visual and physical sensations explanations of the climatic conditions which shape those sensations, using scientific terminology with which the couple are completely unacquainted:

> The sea had the effect of seeming to have a film of glass laid over it, so unbroken was its surface. … Everywhere the black bows of other curraghs peered up mysteriously, looking like the heads of walruses, dugongs, or some such sea-habitants; now visible above the shining surface; now lost to sight; then suddenly reappearing again. It seemed as if they were amusing themselves by some warm-weather game of floating and diving. …
>
> A soft wind blew towards them from the Atlantic, sweeping across their own three islands. You would have thought that, instead of that

48 Ibid., pp. 172–3, 238–9.
49 Ibid., pp. 235–9; Meaney 'Decadence, Degeneration and Revolting Aesthetics', pp. 163–5.
50 *Grania*, pp. 157–8.
51 Ibid., p. 76.

> inhospitable waste of saltness, some region of warmth, fertility and greenness lay out there in the dim and shadowy distance. The air appeared to be filled with soft scents; an all-pervading impression of fertility and growth, strong to headiness, seemed to envelope them as they sat there, one behind the other … The warm air caressed Grania; a sense of vague intoxication and happiness such as she had never before felt seemed to envelope her from head to foot. As it grew darker a quantity of phosphorescence began to play about on the surface, dropping in tiny green rivulets from their oars as they lifted them. It seemed to her as if the queer green glittering stuff was alive, and was winking at her.[52]

Readers who have any scientific or geographical knowledge at all will know that this phenomenon is caused by a warm current from the Gulf Stream and by luminous plankton.

> [N]ature, too was strong; the witchery of the night was strong; the whole combining circumstances of the moment were exceedingly, exceptionally strong. … stopping for a moment in her leisurely rowing, she stretched out her hand and laid it lightly for a moment upon his shoulder …
>
> The intoxicating air had stolen, perhaps, a little into his veins also. And whether spontaneously or merely in mesmeric response to her touch upon his shoulder, he too stopped rowing, and turning a little backwards, rather to his own astonishment put his arm about her waist.
>
> Grania blushed scarlet. Her head swam, but without a moment's hesitation she put her face up to his, and they kissed one another. It was a genuine lover's kiss, their first, although they had been over a year engaged, a fact of which she was immediately and overwhelmingly conscious.[53]

Grania's misdirected passion for Murdough is linked to the warm south, as Algernon Cathers inhabits the Mediterranean. This scene occurs in the chapter preceding Grania's fair-day trauma; whenever Grania tries to come to terms with her feelings about Murdough, both episodes recur to her.[54]

*Grania* contains a vestigial version of the triangle in *Lawrence*, here between Murdough, Grania, and Teige O'Shaughnessy, a hard-working

52 Ibid., pp. 215, 221.
53 Ibid., pp. 222–3.
54 E.g., Ibid., pp. 238, 241, 277.

weaver who loves Grania. Grania takes Teige for granted as a friend though inarticulately sensing that he loves her.[55] Once it occurs to Grania that Teige might be preferable to Murdough as a husband; she realizes there is much to be said for this startling proposition, whereupon her desire for Murdough floods back so strongly she decides she would rather experience the worst with him.[56] A certain inhibition in describing such feelings in a young lady may help to explain why Eleanor Mordaunt's story, which implies similar initial passionate attraction to Algernon Cathers, is told with Lawrence rather than Eleanor as protagonist. During a quarrel with Murdough, Grania startles both of them by suggesting Teige O'Shaughnessy as a better husband;[57] Murdough assumes she is joking, and teases her with the universal ridicule such a match would provoke.[58] There are resemblances between Lawrence and Teige: both lamed and disfigured by childhood falls, both inhabit cabins overlooking the Atlantic and dominated by the sound of its waves.[59] The possibility of a 'happy ending' for Grania with Teige, as for Eleanor with Lawrence, is ruled out because Grania simply does not desire him as she desires Murdough. Teige, for all his virtues, is defeated by Darwinian female sexual selection.

Victorian literary decorum meant Grania's physical desire for Murdough could only be described in symbolic terms. The novel's symbolic dimension goes deeper: it resembles a literary imitation of a Wagnerian opera with the Atlantic instead of the Rhine as the cyclical leitmotif, symbolizing both love and death, from which the story rises and into which it is finally reabsorbed. The novel as a whole might also be read in terms not only of the directionless Darwinian universe but of Wagner's *Tristan and Isolde* (first performed 1865, around the time *Grania* is set) with its Schopenhauerian vision of humanity tormented by instinctive desires which find fulfilment only in acts of will culminating in death. Grania's belief as she drowns that Murdough has come to save her echoes Isolde's dying belief that Tristan has revived, expressed in her final *Liebestod* aria which ends by equating the fulfilment of love with sinking and drowning.

The other partner in Grania's *Liebestod*, however, is not Murdough but Grania's older half-sister Honor, who brought her up and is dying of tuberculosis. Where Grania is an instinctive pagan, Honor is dominated by intense Catholic faith—an official devotional Catholicism admired by the islanders but setting Honor apart from them almost as much as Grania is

55 Ibid., pp. 192–4.
56 Ibid., pp. 252–5.
57 Ibid., pp. 262–4.
58 Ibid., p. 282.
59 A similar room recurs in *The Book of Gilly*. See Hansson, 'Nature, Education and Liberty in *The Book of Gilly*', p. 54. Perhaps a personal memory of Lawless's?

set apart by her passions.[60] Honor possesses quasi-sacerdotal status, regarded as a saint and administering conditional baptism to newborn babies until the non-resident priest can perform the regular ceremony.[61] Honor intensely regrets that she could not be a nun because she had to bring up Grania.[62]

Honor's overwhelmingly sympathetic role should not disguise the fact that Lawless portrays her religious beliefs very ambiguously, as a fantasy of escape shading into worship of death. When Honor rhapsodizes on the glories of the priesthood, Grania recalls being told how a priest beat a child for being late for Mass and says she would have struck the priest. Honor is horrified at this sacrilege: 'Sure a priest isn't a man'.[63] Honor's observation of island life leads her to believe it impossible that women ever derive pleasure from marriage;[64] she dismisses contrary testimony as a delusion inculcated by men.[65] The New Woman novel, to which *Grania* is often assimilated, often ends with the protagonist embracing death to escape unbearable social constrictions, and in this Honor is more similar to Grania—and Murdough—than to other islanders who take their lives for granted.[66] (Grania's reaction to the realization that the defeated woman and her drunken husband represent her probable future is to think that Honor is correct, that there is no happiness in this world.)[67]

Honor's dreams cannot be dismissed as easily as Murdough's, because while Murdough rarely lifts a finger to help others, Honor is tactful and generous in helping her neighbours.[68] She worked to exhaustion all her life for Grania and loves her as intensely as Grania loves Murdough; when she can work no more she laments this burden on Grania,[69] and always puts Grania's interests ahead of her own. Honor intensely fears the torments of purgatory, but would stay there indefinitely to secure Grania's salvation;[70] when Honor is dying and Inishmaan is cut off by fog, Honor pleads with Grania not to risk her life by rowing to fetch the priest, though Honor is convinced dying without the last rites will increase her posthumous sufferings.[71]

60 *Grania*, p. 43.
61 Ibid., pp. 32, 45.
62 Ibid., pp. 131–3.
63 Ibid., pp. 99–102.
64 Ibid., pp. 137–41, 175–7.
65 Ibid., pp. 268–70.
66 For discussion of Grania's death as escape, see C. O'Neill, 'Women, Ambition and the City', 1890–1910', in Pilz and Standlee, *Irish Women's Writing, 1878–1922*, pp. 104–5; Meaney, 'Decadence, Degeneration and Revolting Aesthetics'.
67 *Grania*, p. 238.
68 Ibid., pp. 39–41.
69 Ibid., pp. 96–7.
70 Ibid., pp. 62–4, 134–5.
71 Ibid., pp. 293–6, 317–19.

Grania slips away, leaving Honor attended by a neighbour. She appeals to Murdough, who has been drinking and has a superstitious 'Celtic' fear of evil spirits in the fog;[72] even a passionate embrace which shocks him profoundly fails to secure his assistance.[73] After discovering Teige is away from home, Grania embarks with a small boy. They are wrecked off the Atlantic cliffs of Inishmore. Grania gets the boy onto a rock above high tide mark, but she falls into the ocean, which repeatedly raises her within grasp of safety, then pulls her back:

> Suddenly a quantity of brown seaweed, stirred by the changing tide, swept round the corner of the big rock and floated down towards her. It was a mass of enormously long laminaria, grown, not within tide-marks, but out in the deeper, more abysmal region, as leathery in texture, as solid, and seemingly as sustaining as the branch of a forest tree, the thick strands welded together by years of growth in deep water. It floated up to her, then under her, half lifting her upon itself as upon a rock, her hands clutching in the thick oily strands, her whole body sustained and for the moment uplifted by it. ...
>
> [A]s if Murdough Blake had indeed come to the rescue, and was lifting her in his arms, she let her head fall back upon the seaweed, her cheek resting upon it as if upon his shoulder, her eyes at the same time closing with a long-drawn sigh of satisfaction ... she sank slowly, insensibly, and without a struggle into the great folds of the laminaria, which, after supporting her in that position for perhaps a minute, began gently to loosen its long sashlike strands, floating presently away by degrees over the hardly undulating surface, returning again and again, and sweeping back, though in a less compact mass, now under, now over, now round her, the great brown ribbons swaying in easy serpentine curves about the floating form, the two getting to be hardly distinguishable in the all-pervading dreaminess, a dreaminess of which the very fog itself seemed to be but a part; a dream too deep and apparently too satisfactory to be ever again disturbed or broken in upon by anything from without.[74]

The seaweed, from the deep Atlantic like the warm air which stimulated Grania's attraction to Murdough, is described by its scientific name, which Grania could not know; Grania's final deluded surrender to her passion for Murdough is apparently her final reabsorption by an indifferent Darwinian universe.

72 Ibid., pp. 298–310.
73 Ibid., pp. 311–16.
74 Ibid., pp. 344–5.

Lawless leaves open an alternative, Christian reading. Grania's self-sacrifice does secure a priest for Honor, since the boy is rescued from the rock. The novel ends with Honor dying, consoled by the rites of the church and guarded from knowledge of Grania's death, with a suggestion that Honor and Grania may reunite in the afterlife. Since the Victorian religious novel often culminated in a pious deathbed, Lawless's suggestion could be read as orthodox Christian hope; but Lawless also seems to suggest that just as Grania was absorbed into the sea, Honor is dissolving into the air (filled with Atlantic vapour),[75] and that their reunion involves liberation from passion through dissolution into Schopenhauerian nirvana, symbolized by the Atlantic.

75 Ibid., pp. 347–8.

# 10
# Seumas O'Sullivan and Revivalist Nature Poetry

*Seán Hewitt*

In a mid-nineteenth century review of a collection of poems regarding antiquity, the reviewer turned to make a lengthy aside on the condition of the modern world: 'Are we better than our fathers? We are wiser, but are we happier?'[1] The joy and mystery of the world, according to the reviewer, was rapidly disappearing: 'We have disenchanted all nature. … The meadow which erewhile showed us the fairy ring, as if to put forever to silence all sturdy unbelievers, is now crossed by the massive train-way'.[2] This vision of a specifically Irish 'enchanted' world, contrasted with the industrialized, disenchanted world of modernity, became fundamental, some decades later, to the theorization of the early Irish literary Revival. W. B. Yeats (1865–1939), AE (1867–1935), and many contemporaries sought out fairy-lore, became occult initiates, and cultivated a literature that sought to re-enchant a sterile modern world. This movement, in many ways, was a typical late-Victorian reaction to the loss of orthodox faith, and its nostalgia is symptomatic of contemporary degenerationist narratives, which had doubts regarding the veracity of progressive evolutionism. For many of its writers, encountering Darwin and other evolutionists resulted in a spiritual crisis. In his autobiographical writings, for example, Yeats described being 'deprived by Huxley and Tyndall … of the simple-minded religion of [his] childhood', and claimed to have retaliated by creating 'a new religion' in the form of 'an infallible Church of poetic tradition'.[3] For a number of Revivalists, however, the reaction was quite

1 Quoted in G. Jones, 'Contested Territories: Alfred Cort Haddon, Progressive Evolutionism and Ireland', *History of European Ideas*, xxiv (1998), pp. 199–200. From 'Sketches from the Antique and Other Poems', *Dublin University Magazine*, xxiii (1844), p. 667.

2 Quoted in Jones, 'Contested Territories', pp. 199–200.

3 W. B. Yeats, *Autobiographies: Memories and Reflections* (London, 1995), pp. 115–16.

different. Rather than rejecting science in favour of spiritual exploration, these Revivalists used scientific knowledge as a way of framing their re-enchantment of nature.

One such Revivalist, who still remains a shadowy figure in the history of Irish literature, was Seumas O'Sullivan (1879–1958). Born James Starkey in Dublin in 1879, O'Sullivan wrote eight collections of poetry, a number of pamphlets, a play on the alchemist Nicholas Flamel, and two books of essays. He was an influential publisher, editor of the *Dublin Magazine* (1923–58), and an avid book collector.[4] His father, a pharmaceutical chemist, was also a poet, and contributed to many of the nationalist magazines of his time. Not only this, but he was, in his spare time, a practising field naturalist. In fact, natural history was a shared interest, and both Seumas and his father, William Starkey (1836–1918), contributed to various issues of *The Irish Naturalist* (1892–1924), the organ of the Dublin Naturalists' Field Club, in the late nineteenth and early twentieth centuries. William Starkey noted the appearance of various uncommon butterfly species, including the *vanessa io* ('Peacock') in Rathmines, County Dublin; in a happy coincidence, his son, nearly a decade later, wrote to the *Naturalist* regarding a sighting of the same species in Sligo.[5] Amongst O'Sullivan's papers in Trinity College Dublin, in fact, is a beautifully detailed lepidopterology notebook containing intricate hand-painted illustrations of various moths and butterflies, alongside notes on the Linnaean classification system.[6] Seumas was clearly a talented field naturalist who read the proceedings of the various natural history societies eagerly, and his observations on various species of beetle garnered praise from Robert Lloyd Praeger (1865–1953) and George Carpenter (1865–1939), both eminent scientists and long-term editors of the *Naturalist*. For Seumas, who was particularly interested in beetles, butterflies, and moths, his contributions to *The Irish Naturalist* constituted his first appearances in print, and he had the distinction of having a find recorded in Johnson and Halbert's coleoptera in 1902.[7]

Like his contemporary, J. M. Synge (1871–1909), O'Sullivan's formative years were taken up by various natural history projects, and he worked

4 For more on O'Sullivan's work as editor of the *Dublin Magazine*, see F. Shovlin, *The Irish Literary Periodical, 1923–1958* (Oxford, 2003), pp. 13–38.

5 William Starkey (1836–1918) contributed to the first two issues of the *Irish Naturalist*. See *Irish Naturalist*, i (Nov. 1892), p. 170; ii (June 1893), p. 176. Seumas O'Sullivan (using his birth name, James Starkey) contributed to two later issues. See *Irish Naturalist*, ix (Apr. 1900), p. 108; x (May 1901), p. 115.

6 See Trinity College Dublin, TCD MS 4574. All references to O'Sullivan's manuscripts refer to the Seumas O'Sullivan papers, held at Trinity College Dublin.

7 W. F. Johnson and J. N. Halbert, 'A List of the Beetles of Ireland', *Proceedings of the Royal Irish Academy*, ser. 3, vi (July 1902), pp. 535–827. See J. Russell, *James Starkey/Seumas O'Sullivan: A Critical Biography* (London and Toronto, 1987), p. 25.

regularly with the Dublin Naturalists' Field Club. He read Darwin, Huxley, and works on evolutionism, and his library contained a large number of practical guides to natural history, ranging from the general (such as Lady Kane's Linnaean *Irish Flora* (1833)) to the niche (Anton de Bary's *Comparative Morphology and Biology of the Fungi Mycetozoa and Bacteria* (1887)).[8] This is a testament to both the width and depth of O'Sullivan's interests with regards to natural history, and suggests a man with more than a solely boyhood interest in practical science. O'Sullivan's engagement with naturalist study was coupled with pursuits more traditionally associated with the Irish Revival. A theosophist interested in folklore, his publishing house published books by a number of key Revivalists, and his at-homes were a staple feature of the social life of the Dublin literati. O'Sullivan managed to couple his literary and spiritual pursuits with his scientific interests with little apparent conflict.

Although a keen interest in natural history is not something often studied in criticism of O'Sullivan's Irish contemporaries, it was in fact a surprisingly common feature. W. B. Yeats, Emily Lawless (1845–1913), J. M. Synge, and later Sean O'Casey (1880–1964) were all naturalists before they were writers, though each incorporated or rejected natural history in their own idiosyncratic way. The generalized view of the Revival as an anti-scientific movement is founded on an elision of various engagements with Ireland's burgeoning scientific culture at the close of the nineteenth century. For many Revivalist writers, a study of the natural world that foregrounded its physicality rather than solely its potential for symbolic meaning was an underpinning factor in their poetic innovations. As this chapter will show, Seumas O'Sullivan's rejection of the Celtic Twilight poetry of the 1890s was precipitated by his engagement with a concrete or 'outward' nature, and in his changing aesthetic relationship to symbolist poetics. O'Sullivan, like many of his contemporaries, absorbed and reacted against the materialist narratives of nineteenth-century anthropological and sociological studies, attempting a literary re-enchantment of the natural world. However, instead of turning to invocations of the Celtic gods, or to fairy-lore, O'Sullivan increasingly turned his attention to close study of the physical world, which, over time, was seen as having the ability to enchant without making recourse to the supernatural.

In one of the very few pieces of critical writing on O'Sullivan's work, Sean Mannion has examined O'Sullivan's relationship to material culture in his urban writings, suggesting that the writer moved from a quite typical engagement with Celtic Twilight poetry to a focus, especially in his prose works, on reclaiming the material. For Mannion, O'Sullivan's poems

8 Seumas O'Sullivan papers, TCD MSS 4550–4551.

> abandon the unspoiled pastures and woodlands of Ireland's west for the crumbling cityscapes of Dublin; they focus on the acquisition not of spirit but of detritus such as pins and ragged clothing. And yet, in spite of these startling shifts in emphasis, there exists an essential continuity between these later works and his earlier Celtic Twilight production.[9]

However, O'Sullivan, from relatively early on, began to reconfigure his poetic relationship to the physical world through a recalibration of its relationship to transcendence, attempting to re-enchant the physical in terms which do not conform to typical occult or theosophical mysticism. O'Sullivan's gradual shifting of attention towards the earthly, rather than the supernatural, attempts to de-centre the occult poem by reasserting the materiality of nature.[10]

The symbolist-inspired poetry of the Celtic Twilight reacted against literary naturalism and emphasized the need to capture the non-physical or impressionistic in poetry. For one critic, it is characterized by the 'shadowy and dreamy', and concerned with an 'Otherworld of Druid dreams and fancies'.[11] In his essay 'The Autumn of the Flesh', first published as part of a debate in the literary pages of the Dublin *Daily Express*, W. B. Yeats charted the progress of this Twilight mood in his own verse. The opening section of this essay was adapted from an earlier book review by Yeats of Maurice Maeterlinck's collection of mystical essays *Le trésor des humbles* (1896).[12] Clearly connecting the Irish movement with the work of European contemporaries, Yeats recalls the process via which he arrived at this new poetic style:

> I lost the desire of describing outward things, and found that I took little pleasure in a book unless it was spiritual and unemphatic. I did not then understand that the change was from beyond my own mind, but I understand now that writers are struggling all over Europe, though not often with a philosophic understanding of their struggle, against that picturesque and declamatory way of writing, against that

9 Mannion, 'Celtic Gaslight', p. 47.

10 This reassertion of the material is central to what ecocritic Onno Oerlemans sees as the balance between nature as object and nature as subject. See O. Oerlemans, *Romanticism and the Materiality of Nature* (Toronto, 2002), p. 12.

11 F. Shaw, 'The Celtic Twilight', *Studies: An Irish Quarterly Review*, xxiii (June 1934), p. 260.

12 See W. B. Yeats, 'The Treasure of the Humble', *The Bookman* (July 1897). Reprinted in W. B. Yeats, *Uncollected Prose, II: Later Reviews, Articles and Other Miscellaneous Prose, 1897–1939*, ed. J. P. Frayne and C. Johnson (London, 1979), pp. 45–7.

'externality' which a time of scientific and political thought has brought into literature.[13]

Yeats's reaction against 'scientific thought' and 'externality' is pervasive in his early critical writings, and is posited as the underpinning of his poetry. As John Wilson Foster suggests, 'Yeats believed that Ireland provided the most fertile and hospitable ground available for the counter-Renaissance and the counter-Enlightenment'.[14] Seumas O'Sullivan, however, though he was heavily influenced by the Twilight and symbolist movements, never rejected the evolutionists or the canonical figures of natural history. Rather, he worked at creating a poetry in which the 'external' and 'scientific' could be consonant with spiritual experience. Amongst his library of natural history volumes, there are a number of books on vitalism, alchemy, and physics which suggest O'Sullivan's attempts to grapple with the idea of an animating life-force. One volume of particular interest (and which deals with an idea recently given new credence in biology) is J. M. D. Tupper's *Essay on the Probability of Sensation in Vegetables* (1811), which explores the possibility of 'instinct, sensation and irritability' in flora.[15]

In a way, the Celtic Twilight's disdain for the materiality of the modern world (which was represented on the other side of the *Daily Express* debate by John Eglinton and his call for a poetry of 'the steam-engine and the dynamo ... the kinematograph, phonograph etc.') led to a disdain for materiality in general.[16] John Wilson Foster's rather totalizing characterization of the Revival has been of a movement which turned 'away through language from a more objective world into daydream'.[17] James Joyce, in his article 'The Holy Office' (1904), grouped O'Sullivan alongside his contemporaries in similar terms, dismissing them as a 'mumming company' who 'dream their dreamy dreams'.[18] However, the nature poetry of the Revival era often works to 'effect a reconciliation between science and literature'.[19] Recent critical explorations of this relationship, including Sinéad Garrigan Mattar's *Primitivism, Science,*

13 W. B. Yeats, 'The Autumn of the Flesh', in John Eglinton, W. B. Yeats, A.E., and William Larminie, *Literary Ideals in Ireland* (London, 1899), p. 69.
14 Foster, 'The Culture of Nature', p. 607.
15 James Perchard Tupper, *An Essay on the Probability of Sensation in Vegetables* (London, 1811). See Seumas O'Sullivan papers, TCD MS 4551. For a recent exploration of this idea, see P. Wohlleben, *The Hidden Life of Trees: What They Feel, How They Communicate*, trans. J. Billinghurst (Vancouver, 2016).
16 John Eglinton, 'Mr. Yeats and Popular Poetry', in *Literary Ideals in Ireland*, p. 43.
17 J. W. Foster, *Fictions of the Irish Literary Revival: A Changeling Art* (Syracuse, NY, 1987), p. 181.
18 James Joyce, *The Critical Writings of James Joyce*, ed. R. Ellmann and E. Mason (Ithaca, NY, 1996), p. 151.
19 Foster, 'The Culture of Nature', p. 603.

*and the Irish Revival* (2004), have shown the nuanced influences of nineteenth-century scientific and racial theory on the three original directors of the Abbey Theatre, unsettling the critical commonplace (propounded most vocally by Yeats himself) that the Revival was a generally anti-scientific movement. The influence of science on Revivalism is rarely more prominent than in the writings of those Revivalists who were also practising natural historians, and who moved away from the aesthetic of the Celtic Twilight on terms which reasserted the importance of the physical, 'external' world. As much recent scholarship has shown, Revivalists were engaged in complex ways with the material world they supposedly rejected: many were involved in political, social, technological, and scientific advancements in ways which complicate our understanding of the movement's relationship to the material world.[20]

We can turn to almost any of AE's poetry of the 1890s and early 1900s for an example of how the Twilight mood reads in practice, and how it treats the physicality of the natural world. In 'A Vision of Beauty', from 1896's *The Earth Breath and Other Poems*, AE gives voice to a typical, abstract, 'spiritual' experience of transcending the physical world via occult experience:

> Where we sat at dawn together, while the star-rich heavens shifted,
> We were weaving dreams in silence, suddenly the veil was lifted.
> By a hand of fire awakened, in a moment caught and led
> Upward to the wondrous vision—through the star-mists overhead
> Flare and flaunt of monstrous highlands, on the sapphire coast of night
> Fall the ghostly froth and fingers of the ocean of the light.[21]

The lifting of the veil of the physical world to reveal a visionary, symbolic cosmos is typical of the language of the occult, and is common to Yeats, Fiona Macleod (William Sharp) (1855–1905), and even the early writings of Synge.[22] We can see from this passage how ardently the poem avoids the physical. Each concrete thing is subverted by a qualifier which emphasizes its place in the world of 'essences': the highlands are covered by 'star-mists' which 'flare and flaunt', we get a 'sapphire coast', but this is made abstract ('sapphire coast of night'); likewise, the 'ocean' is not a physical ocean but an 'ocean of the light', which has 'ghostly froth and fingers'. Reviewing this collection for *The Sketch*, Yeats saw AE's work as the apotheosis of Twilight verse:

20 See, for example, P. J. Mathews, *Revival: The Abbey Theatre, Sinn Féin, the Gaelic League and the Co-operative Movement* (Cork, 2003) and P. Reynolds, *Modernism, Drama, and the Audience for Irish Spectacle* (Cambridge, 2007).

21 AE, *The Earth Breath and Other Poems* (London, 1906), p. 14.

22 See, for example, Fiona Macleod, *The Divine Adventure* (Portland, ME, 1903). For Synge, see esp. 'Étude Morbide' (1899) and 'Under Ether' (1897–8), in J. M. Synge, *Collected Works*, II, *Prose*, ed. Alan Price (London, 1966), pp. 25–36, 39–43.

> All things in these elaborate and subtle verses are perpetually changing, and all things are the symbols of things more unsubstantial than themselves. The poet looks at the heavens, and they become a great bird with a blue breast and wings of gold, and at the wood, and it becomes a great sheep shaking its shadowy fleece, and then the bird and sheep become, through some vague wisdom floating in the rhythm and colour of the words, moments of the divine tenderness. ... All things that have shape and weight change perpetually, being, indeed, but symbols. ... 'A.E.' would have us love the invisible beauty before the visible beauty and make our love a dream.[23]

What is valued is AE's mysticism, his insistence on the symbolism of the physical world. Here, Yeats posits AE as the practitioner of an occult art, an art of the initiate, who is able to expand his sense of the 'real' beyond physical nature, transforming the world into a sort of dream. Terry Eagleton, describing Yeats's own work, notes a similar tendency to '[convert] the real to the symbolic', quipping that the poet is often found 'turning a swan into an emblem the instant it glides into view'.[24] As Alex Owen has shown, the fin de siècle occultist and theosophical movements exhibited governing ideas about the control and manipulation of nature, with mystical power concentrated in the power of an enlightened few.[25] In Seumas O'Sullivan's work, by contrast, the move away from Twilight poetics allowed for a reassessment of the place of nature, whereby the physical could be reinstated, and re-enchanted.

Typically, the process of modernization has been seen as being coterminous with disenchantment; and Yeats, in 'The Autumn of the Flesh', clearly positions his view of Twilight poetry in opposition to this desacralization.[26] However, more recently, ecocritical theorists have explored various attempts to 're-enchant' the natural world in response to ecological destruction, and scientific and economic modernity.[27] Neither Yeats nor AE is concerned with the specificities of nature, nor with a sense of enchantment in the material; rather, they emphasize, in line with occultist thought, the lifting of the

23 W. B. Yeats, *Uncollected Prose*, ii. p. 112.

24 T. Eagleton, *Heathcliff and the Great Hunger* (London, 1995), p. 309.

25 A. Owen, *The Place of Enchantment: British Occultism and the Culture of the Modern* (Chicago, 2004), p. 14.

26 'The project of Enlightenment was the disenchantment of the world': M. Horkheimer and T. Adorno, *Dialectic of Enlightenment*, trans. John Cumming (New York, 1972), p. xv. See also Max Weber, 'Science as a Vocation', *From Max Weber: Essays in Sociology*, ed. H. H. Gerth and C. Wright Mills (New York, 2009), pp. 129–58.

27 See, for example, the essays collected in J. Landy and M. Saler, eds., *The Re-enchantment of the World: Secular Magic in a Rational Age* (Stanford, CA, 2009) and G. Levine, *Darwin Loves You: Natural Selection and the Re-Enchantment of the World* (Princeton, NJ, 2006).

veil, the importance of seeing beyond or through the objective, atomistic, 'external' world. For O'Sullivan, a fellow theosophist but also a natural historian, science bolstered, rather than denied, his sense of enchantment in physical nature.

Though AE wrote about O'Sullivan's verse as though it were a continuation of his own project, the younger poet departed significantly from the Twilight model.[28] O'Sullivan's first collection, *The Twilight People*, was published in 1905. O'Sullivan had wanted to call the book 'The Wind among the Hazels', but AE suggested the revised title, perhaps to avoid a too-close similarity with Yeats's *The Wind among the Reeds* (1899), or perhaps to emphasize O'Sullivan as an inheritor of the Twilight aesthetic.[29] In any case, O'Sullivan later regretted the choice. In fact, before the publication of the Boston edition of his *Collected Poems* in 1923, O'Sullivan wrote to the editor, Padraic Colum, instructing him to 'Cut out all that twilight stuff for God's sake. The very thought of it makes me ill'.[30] Even as early as 1912, when putting together a compilation volume of his first three collections, O'Sullivan was already editing a number of the poems from *The Twilight People* out of his oeuvre.[31] However, as Sean Mannion has suggested, the association between O'Sullivan and the Twilight seems to have stuck since the publication of this first collection.[32] O'Sullivan's own friends offered critiques which ranged from the blunt (his poetry lacked 'the hard grasp either of things or ideas') to the constructive.[33] His friend and contemporary Susan Mitchell (1866–1926) offered help which is both telling and fundamental to O'Sullivan's development and his later rejection of the Twilight mode: 'I think this is subjective enough. You must use the terms of external things and images to suggest the inner self—which as yet has hardly a vocabulary of its own—or any scientific terms'.[34] Mitchell recognizes the 'subjective' quality of O'Sullivan's earlier poems and advises an increased focus on 'externality' (that thing which Yeats set out specifically to avoid). Her reference to a need for 'scientific terms' is revealing, suggesting the infiltration of scientific

28 See, for example, AE, 'A Note on Seumas O'Sullivan', *Imaginations and Reveries* (Dublin, 1915), pp. 29–33.

29 P. Kuch, 'Seumas O'Sullivan: Getting Into and Out of Contact', *Hungarian Journal of English and American Studies*, x (Spring/Fall, 2004), p. 169.

30 Russell, *James Starkey/Seumas O'Sullivan*, p. 40.

31 See Seumas O'Sullivan, *Poems* (Dublin, 1912). Excised poems include 'The Sedges', 'The Wild Rose', and 'Two Paths'. Interestingly, O'Sullivan does not choose to present his poems in chronological order of their publication, but instead arranges them in a new order to suggest common themes and progressions.

32 S. Mannion, 'Celtic Gaslight: Urban Material Culture in the Writings of Seumas O'Sullivan', *Éire-Ireland*, xlvi (Spring/Summer 2011), p. 45.

33 Russell, *James Starkey/Seumas O'Sullivan*, p. 49.

34 Quoted ibid., p. 34.

thought into studies of the 'inner self' (i.e., the growing field of psychology) and its implications for O'Sullivan's poetry.

Whereas *The Twilight People* aligns more typically with the Celtic Twilight poetry of the 1890s, *Verses Sacred and Profane* (1908) and *The Earth-Lover and Other Poems* (1909) take on board Mitchell's criticism, moving more closely towards the materiality of the natural world and attempting to view it as both spiritual and spiritualizing. As Ernest Boyd observed in 1916, O'Sullivan's poetry began to take 'a new direction' in 1909, with *The Earth-Lover* especially being 'imbued with a less intangible spirit than its predecessors'.[35] Though Boyd saw AE's selection of emerging contemporary poets, *New Songs* (1904), as the 'manifesto of a school', in which the poets (including O'Sullivan) stood 'in discipleship' to AE, he astutely acknowledged that O'Sullivan's poetry later rejected the 'fairy order of Yeats's early work' and turned to something more physical 'in spite of the glamour of whispering shadows, and evanescent gleams of fairy-land'.[36] The poem 'Communion', from *Verses Sacred and Profane*, acts as an early sort of ars poetica for O'Sullivan's break with the more traditional, ephemeral, non-physical Twilight. Here, the speaker walks out at dusk, as bats 'draw the night on with their wings', finding an enchanted and in some ways animistic natural world which he is both in tune with and in search of: 'I go with will like the wind's will / For solace of earth-exiled things'.[37] Even the title, 'Communion', suggests affinity rather than control.

The idea of animism, central to a number of nineteenth-century studies of 'primitive' culture, pervades O'Sullivan's poetry, and allows for a bridge between the occult philosophy of the theosophists and the physical presence of the atomistic world.[38] As Sinéad Garrigan Mattar has summarized:

> 'animism' was coined by Edward Tylor in his foundational study of 1871, *Primitive Culture*, to describe the first stage in the development of religious thought: the stage at which 'naturalism,' an entirely materialist understanding of the world, was superseded by a stage in which souls and spirits were considered to play a part in the functioning of life. Out of animism was supposed to grow religion proper, ultimately itself to be superseded by scientific rationalism. *Primitive Culture* was a text that was grounded in the idea that culture, like biology, evolved, and

35 Ernest Boyd, *Ireland's Literary Renaissance* ([1916; rev. 1922] Dublin, 1968), p. 259.

36 Ibid., p. 258.

37 Seumas O'Sullivan, *Collected Poems* (Dublin, 1940), p. 59.

38 For canonical discussions of primitive animism, see Edward Tylor, *Primitive Culture: Researches in the Development of Mythology, Philosophy, Religion, Language, Art, and Custom*, I ([1871] London: John Murray, 1920) and James Frazer, *The Golden Bough: A Study in Magic and Religion* ([1890] Ware, 1993).

> that certain cultures (Western ones) had evolved further than others. It depended implicitly on 'the myth of progress.'[39]

The project of re-enchantment in the Revival utilized the concept in various and sometimes contradictory ways. Writers like O'Sullivan and Synge, who both held a continuing interest in science during their literary careers, used the idea to confuse the very concepts of progress and degeneration. In fact, rather than turning to the past for their sources of re-enchantment, both used the principles of scientific positivism to reinstate the material, and thus to understand the natural world as somehow ensouled, or as having a vital agency of its own. The lessons of modern natural history, in other words, were utilized as a way of reinvigorating a disenchanted world without attempting to 'lift the veil'. In O'Sullivan's 'Communion', the speaker is figured as one amongst a number of animated entities; and his movement mirrors those of the elements, which re-enchant in ways which the poet seeks to imitate. This new focus on being subsumed into a non-hierarchical nature is innovative, and performs an act of de-centring, whereby the occult seer no longer looks *beyond* physical nature, but attempts to understand it in a 'pre-modern' mode.[40]

Later, in the poem 'Evening', from *The Earth-Lover and Other Poems*, O'Sullivan gives a more explicit display of his poetics. Here, O'Sullivan, who 'will go out and meet the evening hours / And greet them one by one', engages in a sacralizing of nature which gives an agency to the physical world in creating and controlling its own mystery. O'Sullivan describes a landscape

> Where many a tall poplar summit towers
> On summit, shrines of quietness that send
> Their silence through the blue air like a wreath
> Of sacrificial flame unwavering
> In the deep evening stillness, when no breath
> Sets the faint tendrils floating on light wing
> Over the long dim fields mist-islanded.[41]

The religious language is coupled with a sense of the agency of the natural world, and the poplars are active in the act of sacralizing: they 'send / Their silence' through the air. Though the landscape seems liminal enough, and

39 S. Garrigan Mattar, 'Yeats, Fairies, and the New Animism', *New Literary History*, xliii (Winter 2012), p. 138.

40 Animism, or the vision of an ensouled nature, was posited by anthropologists such as Frazer and Tylor as preceding religion proper, and so a recourse to animistic approaches to nature took on the tenets of a return to 'pre-modern' modes of thought.

41 O'Sullivan, *Collected Poems*, p. 71.

in this way has its root in the Twilight, it does not dissolve away to reveal something sublime and non-physical; rather, its physical presence is key to its holiness. For the majority of the poem, the stillness is uncanny, and animated life seems absent. The lines 'no breath / Sets the faint tendrils floating' refer to the wind, but also give a sense of an unpopulated landscape. This sense of the eerie (aided by the gothic imagery of towers, wreaths, mists) is part of a broader project of re-enchantment through natural science found in the works of Emily Lawless and J. M. Synge, both of whom employ the gothic to recreate the thrilling sense of unease that results from nocturnal pursuits of moths and insects.[42] The evening, conceived as a time when boundaries are fluid, and when the senses are heightened, is a frequent concern for O'Sullivan, who uses it to suggest a nature on the verge of becoming a spiritual presence, a self-sacralizing world. It is this uninterrupted nature that O'Sullivan seeks out. As in 'Communion', the speaker in 'Evening' goes in search of the secrets of the natural world and tries to reinstate them into his vision of physical nature.

Recent critical revisions of 'animism', across a wide range of disciplines, have begun to rehabilitate the term without 'the stigma of belatedness'.[43] Such revisions give us the opportunity to reassess O'Sullivan's particular approach to the natural world, which is both modern and pre-modern. Political theorist Jane Bennett has noted a common critical binary between faith and enlightenment which sets enlightenment disciplines against the notion of an 'enchanted' natural world. As she summarizes, 'Enlightenment is a philosophically informed way of life that builds its home through mastery of a disenchanted world; Faith seeks meaning and comfort through partial reenchantment of the world'.[44] This binary, however, is complicated by writers such as O'Sullivan, who are both avid amateur practitioners of natural science and simultaneously preoccupied with an animated, 'enchanted' vision of nature. Whereas writers like Yeats might use atomistic nature as a symbol for more 'unsubstantial' phenomena, moving against the values of the Enlightenment by privileging that which defies positivist ways of knowing, O'Sullivan's scientific attention to the natural world seeks to re-enchant not by seeing *beyond* the physical, but through an understanding that the physical is, in some way, enchanting. In fact, O'Sullivan often refers to his experience of nature as 'enchanted', and his pursuit of natural history research—with its

42 See, particularly, Lawless's 'An Entomological Adventure', which describes a childhood mothing trip, late at night, drawing on gothic tropes throughout. Emily Lawless, *Traits and Confidences* (London, 1898), pp. 3–35. Synge's 'Autobiography' describes the thrill of mothing in similar terms. See Synge, *Collected Works*, ii. pp. 3–15.

43 Garrigan Mattar, 'Yeats, Fairies, and the New Animism', p. 139.

44 J. Bennett, *Unthinking Faith and Enlightenment: Nature and the State in a Post-Hegelian Era* (New York, 1987), p. 2.

emphasis on close attention, and on the minutiae of species—led easily into this form of wonder.[45]

A number of theorists have noted this sense of enchantment in nature which comes from an increased focus on physical phenomena. Philip Fisher terms it a 'moment of pure presence', while Bennett's definition suggests that it involves 'a condition of exhilaration or acute sensory activity. To be simultaneously transfixed in wonder and transported by sense, to be both caught up and carried away'.[46] J. M. Synge, O'Sullivan's contemporary, tells us of how his own encounter with Darwin and the natural historians led him to be 'more interested in definite life', and this is linked to a sense of the uncanny.[47] Like O'Sullivan, Synge had a profound sense of the numinous in nature, and linked his modern scientific sensibility with that 'of early [man]':

> Natural history did [much] for me ... To wander as I did for years through the dawn of night with every nerve stiff and strained with expectation gives one a singular acquaintance with the essences of the world. The obscure noises of the owls and rabbits, the heavy scent of the hemlock and the flowers of the elder, the silent flight of the moths I was in search of gave me a passionate and receptive mood like that of early [man]. The poet too often lets his intellect draw the curtain of connected thought between him and the glory that is round him. The forces which rid me of theological mysticism reinforced my innate feeling for the profound mysteries of life.[48]

The heightened awareness, the 'passionate and receptive mood', is both connected with natural science and with 'early man'; the figure of the naturalist poet, freed from the obstructive 'curtain' of intellect which Synge associates with other writers, is both modern and pre-modern. The dualism associated with the project of Enlightenment, scientific positivism, and modernity in general is avoided *through* natural history: the attentive poet, newly acquainted with 'the essences of the world', is made alert to 'the glory that is round him'. Synge's encounter with natural science, though made through Darwin, does not result in a perception of his own superior culture and development, but rather emphasizes a connection with more vital, 'primitive' modes of being. This passage reworks the typical associations of materialist science and

45 See Seumas O'Sullivan, 'Prelude to an Autobiography', *Essays and Recollections* (Dublin and Cork, 1944), pp. 83–96.

46 P. Fisher, *Wonder, the Rainbow, and the Aesthetics of Rare Experiences* (Cambridge, MA, 1998), p. 131; J. Bennett, *The Enchantment of Modern Life: Attachments, Crossings, and Ethics* (Princeton, NJ, 2001), p. 5.

47 Synge, *Collected Works*, ii. p. 7.

48 Ibid., ii. pp. 9–10.

Enlightenment with 'the disenchantment of the world' by collapsing any neat vision of progress itself: the acquaintance with natural history becomes an acquaintance with the 'eerie' mystery of nature, and the scientific modernity implicit in being 'rid of theological mysticism' becomes, in its own way, a route to the 'profound mysteries of life'.

O'Sullivan, also interested in lepidoptera, focused similarly on what he termed 'the life of small things', which led him to consider woodlands places of mysterious, spiritual, and biological activity.[49] In a fragmentary draft of a poem in a handwritten notebook, 'some woodland spirit' whispers to the speaker through a fungus, and the 'subtle flutterings' of insect wings give 'faint stirrings' of something 'mythological'.[50] Rather than the figure being a god or embodied deity (as in a later poem by AE, in which the speaker 'walked with a young dryad through the woods'), O'Sullivan's spirit is seen as inhabiting various aspects of the woodland ecosystem.[51] This concept of a spirit within, rather than distinct from, the woodland, emphasizes the immanence of spiritual knowledge in physical nature.

In a number of O'Sullivan's earlier poems, this focus on the physical is given distinct prominence. Throughout his oeuvre, O'Sullivan focused his attention on what he termed 'gleams', or moments of potential transcendence or spirituality experienced in the material world. The first 'gleam' occurs in the second poem of *The Twilight People*, entitled 'The Sheep'. In this poem, which seems startlingly imagistic, the sheep seem to move in an out of time, and the moment at which they connect with the poet's own mind are the moments in which they are said to 'gleam'. Written in one continuous stanza, in which the short lines focus the reader on the slowed pace and the insistence on the specificity and simplicity of each image, the poem begins with a concrete description of the scene:

> Slowly they pass
> In the grey of the evening
> Over the wet road,
> A flock of sheep.
> Slowly they wend
> In the grey of evening
> Over the wet road
> That winds through the town.

49 TCD MS 4560, fo. 61. From the evidence in this notebook, I would suggest that this poem was written before 1902 (i.e., during the time that the poems in *The Twilight People* (1905) were being written).

50 TCD MS 4560, fo. 61.

51 AE, 'Platonics', *Vale and Other Poems* (London, 1931), p. 23.

Slowly they pass,
And gleaming whitely
Vanish away in the grey of the evening.[52]

These sheep then become, through repeated imagery and language, akin to the memories of a relationship, which 'Loom for a moment, / Gleam for a moment'. Most revealing, here, is the way in which the memories and the sheep are given equal precedence in the structure of the poem, with both being repeatedly equated to the point at which, in the final lines, it is difficult to ascertain which is being referred to:

We two together,
Went with slow feet
In the grey of the evening
Where the sheep lay.
Whitely they gleam
For a moment and vanish
Away in the dimness
Of sorrowful years,
Gleam for a moment
All white, and go fading
Away in the greyness
Of sundering years.[53]

The poem opens with the speaker recounting a present-tense scene, in which sheep are moving down a road. By the time we get to this final section, the conflation of the sheep with surfacing memories is almost complete, and is enacted by a simple full-stop. By ending the sentence after 'Where the sheep lay', O'Sullivan leaves the reader to decide whether the following (final) sentence returns to the present or remains in the realm of recollection: 'Whitely they gleam' could just as well be referring to the sheep as to the memories, and thus the remaining imagery achieves a satisfying subtlety. The moment at which the physical becomes metaphorical, and vice versa, is made more difficult to identify: O'Sullivan's 'gleams' are moments during which the physical and transcendent are coexistent. Rather than the physical disintegrating in favour of non-physical phenomena, the 'outward' is given equal value.

In the untitled prefatory poem to *Poems* (1912), O'Sullivan contextualizes his poetry from the previous seven years in this context of re-enchantment, emphasizing the potential of modern writers, writing from modern experience,

52 Seumas O'Sullivan, *The Twilight People* (Dublin, 1905), p. 2.
53 Ibid., p. 3.

to create something which might rival 'that fierce olden ecstasy' to produce 'a wisdom winnowed from light words'.[54] Although O'Sullivan laments that the modern reader and writer have come 'late in a too travelled world', drawing on typical Revivalist nostalgia and degenerationist narratives, he suggests that perhaps 'our sadder hearts / Have dealt in more enchanted marts / Than those old singers, and our eyes / Have gathered costlier merchandise'.[55] Setting the poem within this world of increasingly market-driven economics, the poet carves out a route for an enchantment of modern life in lines with a more progressive ethos, finding in the material world things to rival the 'magic of wood and wind and wave' that constitutes the fabric of bardic poetry. In this way, O'Sullivan's compilation of his own poetry is framed as an act of modern re-enchantment, an attempt (like the Twilight poetry of the 1880s and 1890s) to recover the spiritual, but coupled with the theory that this might be done *through*, rather than apart from, what Yeats would call the 'externality' of a 'scientific' age.

The new poems added by O'Sullivan to this selection confirm this reading, and mark a decisive break from the Twilight aesthetics which had crept into his early verse. 'The Gleam', referring back to O'Sullivan's theory of momentary transcendence outlined in 'The Sheep', is the most comprehensive statement of this new poetic approach. It begins with an imperative, ardently refocusing the reader's attention away from the Celtic gods popular with the Twilight poets and towards the mysteries of physical nature. From there, O'Sullivan deconstructs the two types of 'wisdom' available through magic and through a study of nature, concluding that we should look for enchantment, spiritual experience, and knowledge not in the supernatural, but in the workings of the physical world:

> Bundle the gods away:
> Richer than Danaan gold,
> The whisper of leaves in the rain,
> The secrets the wet hills hold.
> To follow a leaf on the wind,
> And joy in the following:
> Yea, in the following find
> More than the wise man found,
> Merlin the foolish wise
> Who followed on magic ground
> A vision beyond the eyes
> Of any of mortal born;

54 O'Sullivan, *Poems*, p. 2.
55 Ibid.

Wisdom the highest this,
Mortal so shalt thou live,
And living heed not the scorn
Of the gods in their lonely bliss,
Therefore guard well thy heart.
O earth-born, harbour thou there
No vision but earth can give
No rapture but earth may share.[56]

Despite the clunky archaisms, O'Sullivan's poem chimes with another famous poetic rejection of the Celtic Twilight. J. M. Synge's 'The Passing of the Shee', which carries the subtitle 'after looking at one of AE's pictures', asserts its difference from the old aesthetic and, like O'Sullivan's 'The Gleam', refocuses the poetic vision earthwards.

Adieu, sweet Angus, Maeve and Fand,
Ye plumed yet skinny Shee,
That poets played with hand in hand
To learn their ecstasy.

We'll search in Red Dan Sally's ditch,
And drink in Tubber fair,
Or poach with Red Dan Philly's bitch
The badger and the hare.[57]

Again, as with O'Sullivan's poem, 'The Passing of the Shee' acts as an ars poetica for Synge's attempts at a more 'brutal' poetry that has 'strong roots among the clay and worms'.[58] Though 'The Gleam' does not propose a 'brutal' poetry, it does reject the Celtic gods in favour of the physical world. Synge's 'plumed yet skinny Shee' suggests the decorative, rather than substantive, quality of the supernatural in the Twilight aesthetic, and O'Sullivan's later rejection of the supernatural suggests a higher wisdom in the pursuit of nature study than in the pursuit of mystical or religious activity. Two years later, in his poem 'Credo', O'Sullivan subverts the Christian creed in favour of a nature-worship based in an animistic world:

I cannot pray, as Christians use to pray,
Before the holy Rood,

56 Ibid., p. 41.
57 J. M. Synge, *Collected Works*, I, *Poems*, ed. Robin Skelton (Gerrards Cross, 1982), p. 38.
58 Ibid., p. xxxvi.

Nor on the sacred mysteries seven as they
Believing brood. [...]

For I [...]
Have known the sun, the wind's sweet agency
And the soft rains that bless
And lead the year through coloured pageantry
To fruitfulness.[59]

Throughout O'Sullivan's poetry and prose, there is a strong association between a sequestered Christian orthodoxy which is wilfully separated from the natural world and an attempt to recover the spiritual in nature.[60] Unlike Yeats and AE, who look beyond nature and beyond 'externality', O'Sullivan looks at the material world as a symbol of its own spirituality and mystery. The wind has a 'sweet agency', the rains 'bless' the land: this is an auto-sacralizing world, and is part of a poetic project which increasingly attempts to bring the spiritual back into the material. For O'Sullivan, there should be 'no rapture but earth may share'. The gods must be 'bundled away' in favour of an intimate knowledge of nature which is itself spiritual.

O'Sullivan's sacralizing of nature extends, in his later urban poems and the essays in *Mud and Purple* (1918), into a self-reflexive aesthetic which at once seems to satirize and to defend the poet's often hallucinatory experiences of spirituality in the material world. In his essay 'The Georgian (Dublin)', O'Sullivan repeatedly witnesses the apparition of a figure on the Dublin streets. The essay discusses the modernization of the urban landscape, where the nineteenth century is dismantled and replaced by modern architecture. As one building is removed and replaced with another, O'Sullivan sees the march of progress at work: 'The last scaffolding had been removed. The ancient had vanished before the modern; the new century had asserted itself, and with its coming came all the noise and turmoil and unrestful colour of the more progressive times'.[61] This modernization results specifically in a disenchantment, as the apparition O'Sullivan has witnessed disappears. At certain times, there are 'gleams' of the old world, and the whole process leads to a self-reflexive assessment of the poet's acts of enchantment. On one occasion, 'the street lamps were shining with the faint glow of deep-sea flowers, and in their uncertain and transmuting glow I seemed to see the street again as I had

59 O'Sullivan, *Collected Poems*, p. 131.

60 See, for example, his poem 'The Convent', ibid., 211 and his essay 'In the Botanic Gardens', *Mud and Purple: Pages from the Diary of a Dublin Man* (2nd edn., Dublin, 1918).

61 O'Sullivan, *Mud and Purple*, pp. 25–6.

known it two years ago'.[62] The material world transfigures itself into something which promotes mystical experience; however, by the end of the essay, the role of the poet within his own text is brought into question. O'Sullivan reflects on the disappearance of the ghost figure, who represents the 'lost' century:

> Had I ever really seen him? Was it possible that my daily sojourning amongst these shadowy streets and lane-ways had excited my imagination into the creation of this dweller in their midst? ... Had I not, too, seen in these ancient doorways strange, life-like shapes and figures which had vanished on my approach? Had I not, on more than one evening, seen the great windows of tall houses blazing with myriad candles, and heard the roisterous companions within? Had I not even tried to respond to their chorusing, and found myself shouting foolishly before the lifeless doorway of some deserted tenement?[63]

This is an evocative and revealing image for O'Sullivan as a poet and re-enchanter. His project of recalling the spiritual to the material as an act of imagination is brought into question; his engagement with the past, and the processes of both urban and rural modernization, lead to a sort of hallucinatory experience whereby the separation of past and present, of spiritual and physical in his writing, becomes potentially divorced from the reality of the modern world. The figure of O'Sullivan 'shouting foolishly' at an empty tenement is both a resigned acknowledgement of his delusion and an apt metaphor for his poetic project, which finds itself constantly concerned with the material, with the 'external' things to which he attempts to recall a sense of haunted spirituality.

The religious quality of O'Sullivan's poetry was understood early on. *Dana*'s review of *The Twilight People* noted a sort of 'hushed reverence', and saw the poems as an attempt to put into language the 'inexpressible rapture of the sense of unity of the adorer and the adored'.[64] However, as we move through O'Sullivan's career as a poet, the location of the spiritual element is moved closer into the physical world, and the potential of imaginative or mystical transcendence brought into question as a literary aim. Over the turn of the nineteenth century and into the twentieth, the younger generation of poets began to re-materialize poetry, but also sought to re-sacralize the material. Concern with modernization and its concomitant disenchantment leads, in

62 Ibid., p. 26.
63 Ibid., pp. 26–7.
64 "T.K.", 'A Literary Causerie', *Dana*, xi (Mar. 1905). O'Sullivan's poems were published often in the pages of *Dana* (1905–6), a short-lived Dublin magazine edited by W. K. Magee (1868–1961) and Fred Ryan (1876–1913).

O'Sullivan's poetry and prose, to a self-conscious quest for a poetry which enshrines the spiritual within the physical world of nature. As O'Sullivan's work shows, the Revival's engagement with materiality was on occasion an attempt to recalibrate the relationship between the material and spiritual in regard to the natural, bringing the 'gleams' of spiritual transcendence closer to the physical world.

# Index